Electronic Projec[t]

Other Constructors Projects Books

Electronic Projects in Audio
Electronic Projects in the Car
Electronic Projects in the Home
Electronic Projects in the Workshop
Projects in Radio and Electronics
Electronic Game Projects
Electronic Projects in Music

Electronic Projects in Hobbies

F. G. Rayer
Series Editor Philip Chapman

Newnes Technical Books

The Butterworth Group

United Kingdom   Butterworth & Co (Publishers) Ltd
London: 88 Kingsway, WC2B 6AB

Australia   Butterworths Pty Ltd
Sydney: 586 Pacific Highway, Chatswood, NSW 2067
Also at Melbourne, Brisbane, Adelaide and Perth

Canada   Butterworth & Co (Canada) Ltd
Toronto: 2265 Midland Avenue, Scarborough,
Ontario M1P 4S1

New Zealand   Butterworths of New Zealand Ltd
Wellington: T & W Young Building
77–85 Customhouse Quay, 1, CPO Box 472

South Africa   Butterworth & Co (South Africa) (Pty) Ltd
Durban: 152–154 Gale Street

USA   Butterworth (Publishers) Inc
Boston: 19 Cummings Park, Woburn, Mass. 01801

First published 1979

© Butterworth & Co (Publishers) Ltd, 1979

All rights reserved. No part of this publication may be reproduced or transmitted in any form or by any means, including photocopying and recording, without the written permission of the copyright holder, application for which should be addressed to the Publishers. Such written permission must also be obtained before any part of this publication is stored in a retrieval system of any nature.

This book is sold subject to the Standard Conditions of Sale of Net Books and may not be re-sold in the UK below the net price given by the Publishers in their current price list.

British Library Cataloguing in Publication Data

Rayer, Francis George
   Electronic projects in hobbies.
   1. Electronic apparatus and appliances —
   Amateurs' manuals
   I. Title
   621.381   TK9965   78-40662

ISBN 0-408-00354-5

Typeset by Butterworths Litho Preparation Department

Printed in England by William Clowes & Sons Limited
Beccles and London

Preface

Electronics now plays a large part in hobby activities of many kinds. Electronic means are popular where other methods were once employed, and these new techniques will be found in projects for electric trains and other motor driven models, giving improved running. For the photographer, an enlarging exposure meter and process timer will be useful. Other items of immediate interest include a treasure or metal locator, a Morse code oscillator practice set, a metronome for the musician, and a home intercom. Uses also in the home will be found for a freeze indicator and light operated alarm. For those looking primarily for amusement, a manual control test and a reaction indicator are included.

Projects range from very easy to quite complex, and the aim throughout is to provide both an understanding of how a device operates, and a working unit which can be relied upon to function well. There is thus a two-fold reward to be obtained from each project.

Contents

Introduction

1 Manual control tester   1
2 Model train 'accelerate/decelerate' control   6
3 Flasher for models   10
4 Morse code practice oscillator   14
5 Unijunction transistor metronome   19
6 Enlarging exposure meter   26
7 'Freeze' indicator   32
8 Adjustable power supply unit   38
9 Light actuated receiver and lamp unit   44
10 Home intercom   53
11 Metal locator   61
12 Process timer   71
13 Reaction indicator   79

Appendix   87

Introduction

With constructional projects such as these it is worth noting that there is generally reasonable latitude in values, as this will simplify obtaining components. With electrolytic capacitors, as example, there is considerable tolerance, and 4.7µF, 5µF or 6.4µF values are virtually interchangeable, as also 20µF, 22µF or 25µF, or 100µF and 125µF, and so on. The working voltages of such components vary greatly, so choose one at least equal to the voltage present.

Capacitor values may be shown in picofarads (pF), nanofarads (nF) or microfarads (µF). Reference to the following will avoid confusion

$$10\,000\text{pF} = 1\text{nF} = 0.001\mu\text{F}$$
$$10\,000\text{pF} = 10\text{nF} = 0.01\mu\text{F}$$
$$100\text{nF} = 0.1\mu\text{F}$$

Thus 4700pF, 4.7nF or 0.0047µF capacitors would all be of the same value. For reasons similar to those above, 0.005µF could be substituted. Exact values as specified are generally necessary only in resonant circuits, such as the metal locator.

In a similar manner, resistor values are usually not critical. However, the correct values are readily available, though items such as 20kΩ, 22kΩ or 25kΩ potentiometers could replace each other, or 200kΩ and 220kΩ be regarded as interchangeable.

With transistors, it is impossible to list the numerous equivalents and similar types which could be incorporated in a circuit. Types specified are readily available, and where alternatives are adopted for any reason, care should be taken that they have characteristics closely similar to those indicated.

The projects described in general need cases, for protection and appearance. However, there is often considerable latitude here, and sometimes no need for a case exactly as illustrated. For some items, the constructor may prefer to make a wooden case; others can be housed in inexpensive plastic lunch boxes, and similar household or general purpose containers. Various metal and insulated boxes and cases are available from suppliers of electronic components. The 'universal chassis' type of case (Home Radio, 240 London Road, Mitcham, Surrey CR4 3HD) is available in many sizes, and is inexpensive. Vero Electronics (Industrial Estate, Chandler's Ford, Hants. SO5 3ZR) make many cases of professional finish. Bare metal cases or boxes can be painted, sprayed, or covered with self-adhesive material.

Naturally construction will be most straightforward when the board layouts are followed exactly as shown, but there would often be some latitude in the size of board, and layout.

To avoid any difficulty in following board layouts exactly, boards should be cut so that the number of rows of holes illustrated will be available on both dimensions, with foil conductors as shown.

1

Manual Control Tester

This type of device always provides amusement and tests the manual dexterity of the person attempting to operate it. There is a 'course' made from a shaped rod or stout wire, and the competitor has to pass a small ring along this, without allowing the two to touch (Fig. 1.1). It

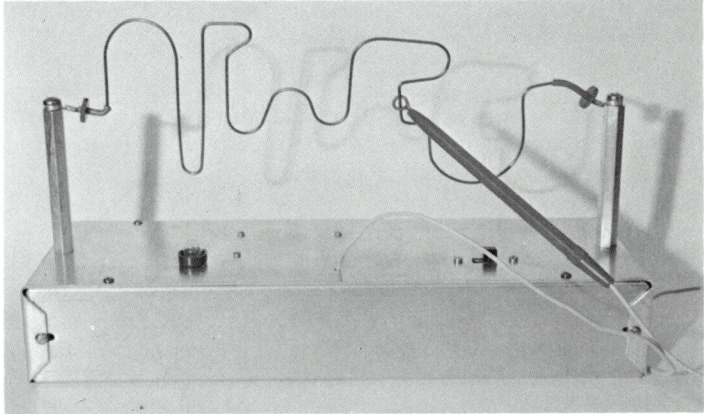

Figure 1.1
The manual control tester

is easy to arrange for any degree of difficulty, either by changing the size of the ring, or by setting a time limit on completion of the course. Failure — that is, contact between ring and rod — is shown by a bulb which lights up and remains lit even if the contact was only momentary.

The circuit

A thyristor or silicon controlled rectifier (SCR) is used as the latch-on device (Fig. 1.2). Normally, the SCR provides no conduction from its cathode K to anode A. With battery switch S1 closed, no current flows through the SCR or bulb, so the latter is not lit. If contact is made between the ring and rod, current from battery positive flows through the lamp and R2 to the SCR gate G. This triggers the SCR into conduction, and it remains in this condition even if the gate voltage is

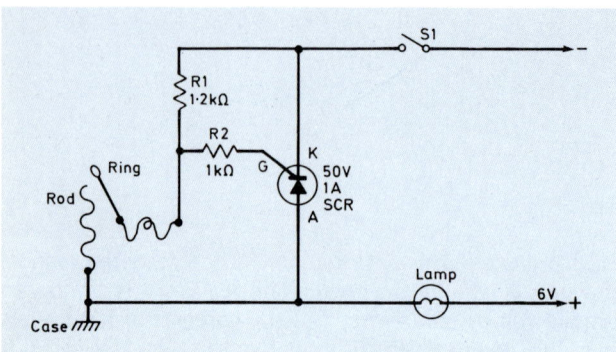

Figure 1.2
Contact between ring and rod triggers SCR into conduction

removed. The lamp thus stays on, showing that the test has been failed. To extinguish the lamp it is necessary to open switch S1 briefly, and the SCR then returns to its non-conducting state.

Resistor R1 holds the gate circuit negative, even with some slight leakage between rod and ring circuits, such as from dampness or finger contact, which can in some circumstances trigger a sensitive SCR. R2 limits the gate current when rod and ring are in contact with each other.

It is convenient to use a 6V battery supply, with a 6.3V bulb, though this voltage is not essential. The lamp voltage should be approximately the same as the voltage of the battery.

It would be possible to connect a bell, buzzer, or audio oscillator instead of the bulb, though the bulb is simpler and has advantages where noise may be a nuisance to others. A buzzer or bell will have an interrupter and will thus allow the SCR to return to non-conduction, so a resistor of about 470Ω should be placed in parallel with the buzzer terminals, to provide a current to hold the SCR on.

Construction

The resistors and SCR can be placed as shown in Fig. 1.3. It is convenient to fit red and black flex for positive and negative (S1), with some other colour for the ring connection, which should be 50cm or so long. The bulb holder can be fully insulated, or can have its outer part earthed to the chassis, the centre contact tag running to battery positive.

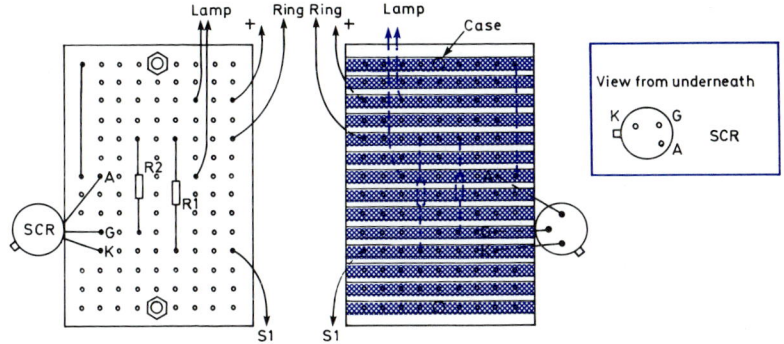

Figure 1.3
Both sides of the perforated board

Note that the SCR anode A is returned to the case or chassis, by means of one of the bolts which secure the board. These bolts pass down through the chassis, and extra nuts can be used to give a little space between board and chassis.

Chassis and rod

A chassis 30 × 10 × 5cm allows a long enough rod to be fitted. It is supported by threaded pillars (Fig. 1.4) and is thus in contact with the chassis. The rod used was 60cm long and about 2mm thick, and it was obtained from a supplier of materials for model engineering. It could be 1mm thick or very stout, bare wire. A suggested shape is shown but it could also have loops backwards and forwards.

About 1cm of insulated sleeving, and an insulated washer, are put on each end, to give a place to rest the ring. The ends of the rod should be soldered to tags, or formed into loops, and securely held by screws.

The ring is also made from stout wire, formed into a circle about 5mm in diameter. The wire runs down the body of a used ball-point pen, to serve as a handle, and is held inside by adhesive or one or two matchsticks, if necessary. The pen top is drilled for the flex, and is screwed in place after soldering the flex and wire together. After placing the ring on the rod, solder the ring so that it cannot open.

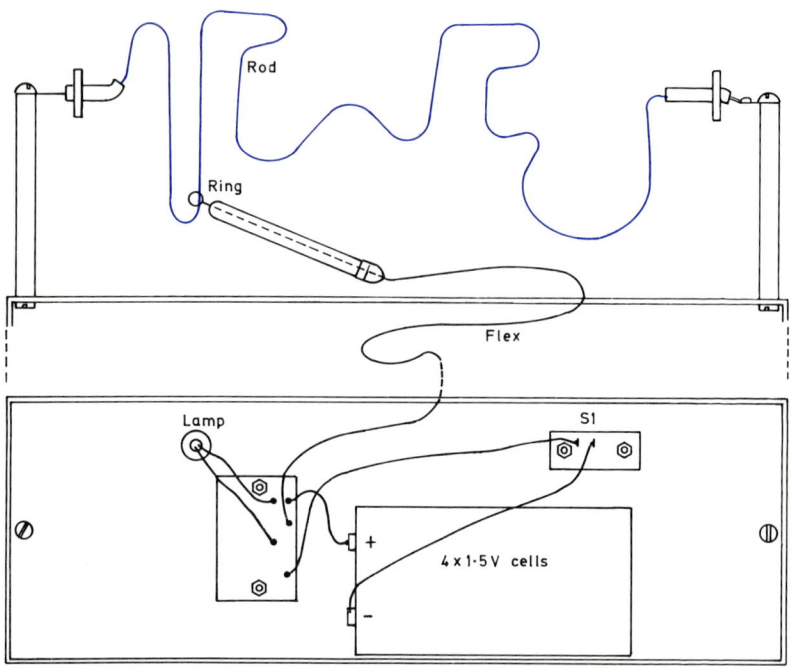
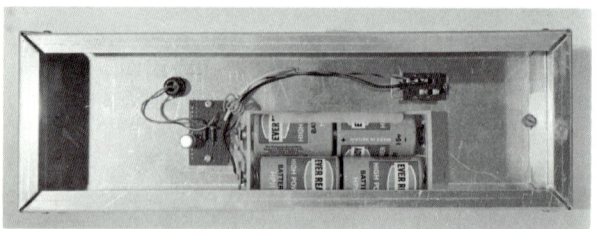

Figure 1.4

Details of assembly. The shaped rod is electrically common to the metal case

Using the tester

With S1 closed, momentary contact between ring and rod should cause the bulb to light until S1 is opened. No component values are critical and almost any SCR is likely to operate (the lower voltage types are inexpensive). Battery polarity must be as shown, with the SCR anode positive.

Circuit checks

A fault is very unlikely, but if the device does not operate, check that there is contact from the SCR anode to chassis and the shaped rod.

Table 1.1. Components list for manual control tester

Resistors	
R1	1.2kΩ (5%, ¼W)
R2	1kΩ (5%, ¼W)
Semiconductor	
SCR	50V, 1A or similar
Miscellaneous	
Bulb	6.3V, 0.15A with holder
S1	*on-off* switch
	perforated board about 25 × 35mm
	box chassis, 30 × 10 × 5cm
	two 2 B.A. or 4 B.A. pillars 8cm long
	materials for course and ring as described
	flex, battery holders etc

Should it be necessary, the switch S1 can be tested by shorting it, and the bulb can be checked by means of the battery. If the lamp still fails to light when the junction of R1 and R2 is connected to battery positive or SCR anode, check R2 and the SCR.

Four 1.5V cells in a holder provide a convenient supply, so it is necessary to make sure these are inserted the right way, to obtain the polarity shown.

2

Model Train 'Accelerate/Decelerate' Control

With a model train, switching the current on starts the engine with an unrealistic rush, while switching off results in an abrupt stop. The control unit shown in Fig. 2.1 avoids this. Current and voltage to the engine are electronically controlled so that there is a rise in power to start the train and bring it up to speed, or a slow falling away of power when stopping so that the engine slows realistically to a halt.

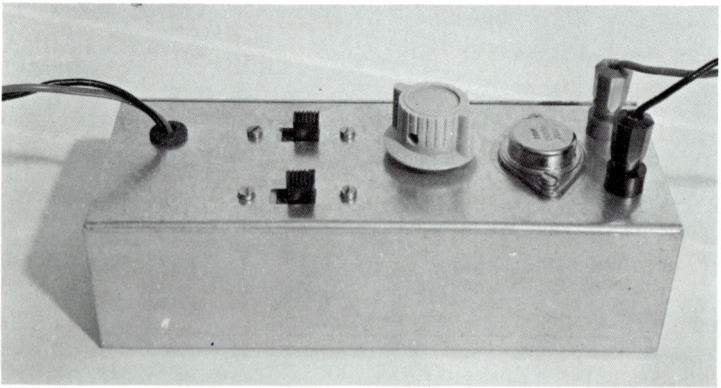

Figure 2.1
The model train controller

Similar effects can be obtained with a manually operated speed controller, but the unit shown here accomplishes this automatically. It may also be used with forms of automatic control, where a section of track is switched in or out of circuit in circumstances where a manual speed controller will not be present.

The circuit

The circuit is shown in Fig. 2.2 and it has two switches, S1 and S2. S1 is for complete on-off switching, taking power off the unit when it is not in use.

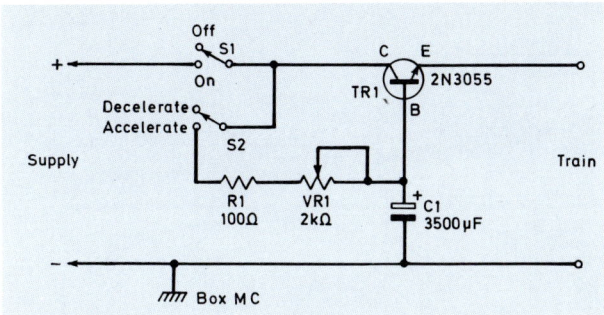

Figure 2.2
Charging and discharging of C1 controls the current through TR1

S2 has two positions. When S2 is in the decelerate position, and the model has been running, the engine will slow to a halt. Moving S2 to this position is thus the equivalent of 'stop' except that power disappears gradually. Placing S2 in the accelerate position is similarly equivalent to 'start' except that power builds up slowly, not being abruptly applied as with a straightforward on-off switch.

Delay and control are obtained by VR1, C1 and TR1. If S2 is open and the engine halted, C1 is discharged so that TR1 does not conduct. When S2 is closed, C1 charges through R1 and VR1, so there is a steadily increasing positive bias for the base of TR1, followed by a rising emitter current for the model. VR1 is pre-set, and controls the 'acceleration' or rapidity with which current builds up.

With S2 moved to the open position, TR1 continues to pass current, this falling away as C1 discharges.

Construction

Fig. 2.3 shows all wiring and components; a metal box about 15cm × 6cm and 4cm deep is convenient. Slots for the slide switches can be made by drilling a few holes close together, then opening out with a small flat file. The supply is obtained from red and black flexible leads, which pass through a grommet. Also drill for the potentiometer VR1, output sockets, and TR1.

When mounting TR1, note that an insulation set is used. This consists of a thin mica washer which is placed between the transistor and panel, and bushes to put on the fixing screws. Base B and emitter E pins pass through clearance holes, and all burr must be removed from the metal before fitting the transistor. A tag under one nut provides the collector connection C.

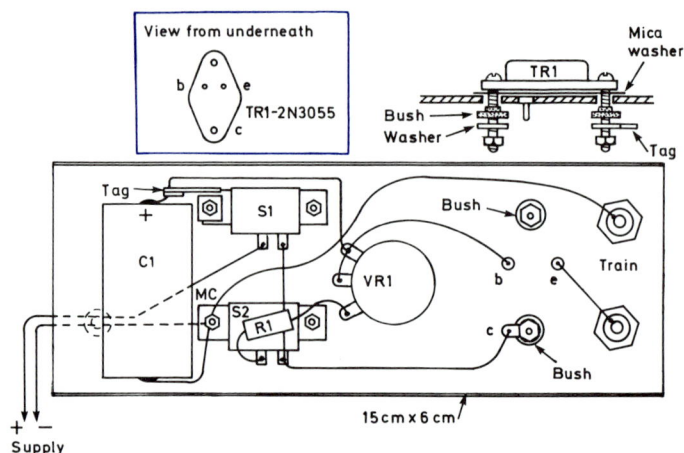

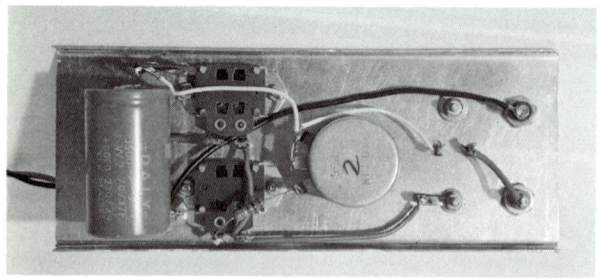

Figure 2.3

Assembly in metal box, and details of TR1 mounting

The large capacitor C1 is supported at its positive end by the insulated tag. The short tag strip is held by one of the nuts fixing S1.

Current for the unit can be supplied by a battery or mains unit giving the usual d.c. output. The polarity of supply must be as shown. Any reversing switch thus has to be placed between the control unit and track.

Using the controller

VR1 is adjusted to give the required rate of speeding up, or acceleration. With some engines the slowest rate of acceleration (VR1 wholly in circuit) may result in the train not starting at all. This could be avoided by reducing VR1 to 1kΩ, but then the slowest start which can be achieved by adjusting a 2kΩ potentiometer is not available, so on balance the larger value is probably better.

Table 2.1. Components list for model train accelerate/decelerate control

Resistors	
R1	100Ω (¼W)
VR1	2kΩ linear potentiometer
Semiconductor	
TR1	2N3055 (with insulation set)
Miscellaneous	
S1 and S2	slide *on-off* switch
C1	3500μF, 15V electrolytic
	metal box about 15 × 6cm, 4cm deep; tagstrip, output sockets etc

The unit operates with the popular layouts which require up to 12V or so maximum. As with other methods of slow starting, the track should be clean, and engines should have proper contact and run freely.

As the circuit is adjustable, values are not too critical. Smaller values for C1 will reduce the accelerate/decelerate delay, while larger values will increase it.

3

Flasher for Models

This unit will flash two lamps in an alternating sequence, or can flash on and off a single lamp when this is more suitable. Two lamps are appropriate for certain types of crossings, and are also effective to get attention in some other applications, rather than the use of a constantly illuminated indicator lamp. The single lamp is correct for some models, and can provide about equal periods lit or extinguished. It may also be fitted in this form in electronic equipment where the flashing signal is preferred.

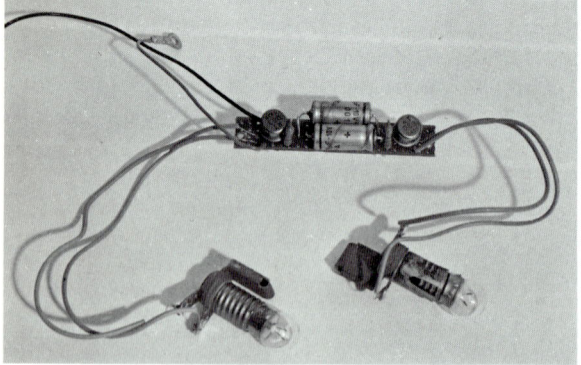

Figure 3.1

The unit can be assembled on a small board only 6 cm x 1 cm

Compact construction (Fig. 3.1) allows the unit to be accommodated in small models, and leads running from it permit the lamps themselves to be situated where required. The low current drain means it can usually be run from the same battery or other supply employed for the model or other equipment.

The circuit

The circuit (Fig. 3.2) is a multivibrator operating with a frequency of about 1Hz (one cycle per second). Lamps L1 and L2 pass the collector current of TR1 and TR2, so flash on and off alternately. The actual frequency can, if wished, be altered between wide limits, by changing

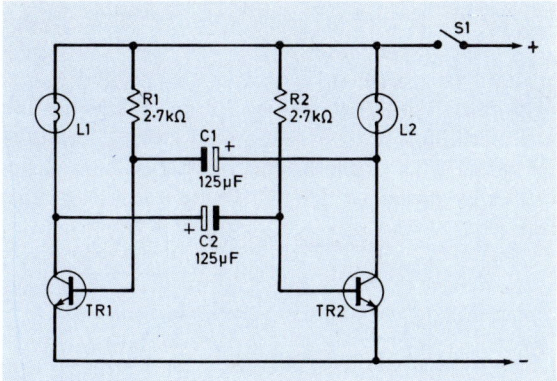

Figure 3.2
Multivibrator with lamps acting as collector loads

resistor and capacitor values. Where L1 and L2, TR1 and TR2, R1 and R2, and C1 and C2 are similar, the 'on' and 'off' periods of the lamps will be about equal.

When the multivibrator is running, TR1 or TR2 may be conducting. The circuit is not stable in either of these conditions, so alternates between them, or is free running. Frequency is largely determined by the resistor and capacitor values.

When S1 is closed or power is applied, lack of perfectly symmetrical values will mean that one transistor conducts slightly more than the other. Assuming that TR1 passes the larger current, there is a slightly greater current through L1 than through L2. As a result of the voltage drop in these lamps, the collector potential of TR1 will be lower than that of TR2. C2 thus starts to swing TR2 base negative, further reducing TR2 collector current. The voltage drop in L2 is thus reduced, and C1 maintains TR1 base positive, so that TR1 conducts heavily. The result is that TR1 is passing the full current available through lamp L1, while TR2 is not conducting, so that L2 is extinguished.

In this condition (with TR1 fully conducting) TR1 collector potential is nearly zero. C2 has not charged, so TR2 base is also nearly fully negative. However, C2 is charging through R2, so that the base of TR2 is moving positive. As collector current through L2 rises, TR2 collector potential is moved negative, and TR1 base is swung negative by C1, so that TR1 is in turn cut off. The multivibrator alternates between these two conditions — TR1 conducting and TR2 cut off, then TR2 conducting and TR1 cut off. This continues, repetition rate depending on the circuit constants.

Construction

The components are fitted to a small 0.1in matrix board about 6cm × 1cm. C1 and C2 will project slightly, as in Fig. 3.3 but the whole unit will fit in a very small space. Foil breaks are required under C1 and C2, and are also provided at the ends of the board to avoid possible shorts when mounting the unit by means of the two fixing holes. Note that C1 and C2 leads cross over.

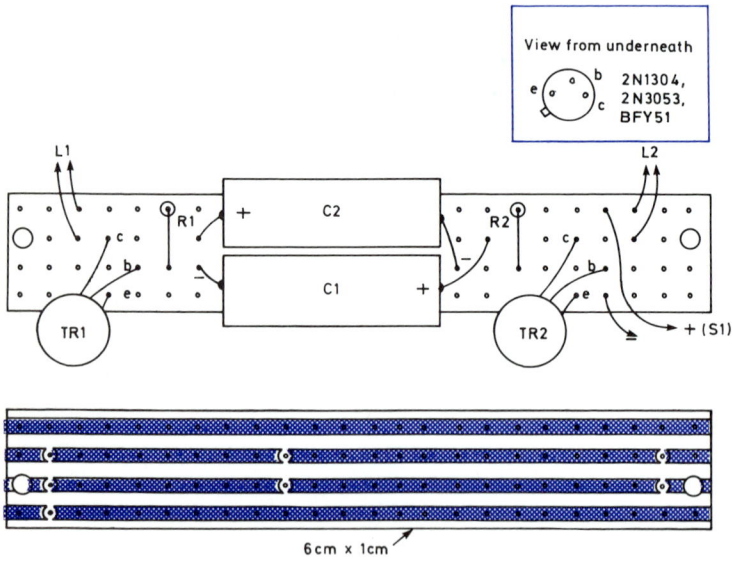

Figure 3.3
Positioning of components on the board

Six flying leads of thin flex are provided, to run to L1, L2, and the supply circuit. The latter should be colour coded red and black for positive and negative, as correct polarity is necessary.

The transistors are near the board, leads being shown longer than necessary for clarity. R1 and R2 leads, or other bare conductors, must not touch the transistor cases.

Using the flasher

The circuit operates from about 6V. It may run from 12V, the 6V lamps being retained, if a resistor of about 100Ω (1 watt) is placed in one supply lead.

If only one flashing lamp is required, replace one lamp by a 120Ω ½W resistor.

The unit can be operated from a.c., or from circuits where the polarity may be changed by a reversing switch, by adding four 1N4001 or similar rectifiers, or using a small low voltage bridge rectifier to provide the polarity shown (emitters to negative).

Faults are not very likely. If the flasher does not operate, check that foil breaks have been provided under C1 and C2, and that no shorts between adjacent foils arise from fragments of solder, or the method of fixing.

Table 3.1. Components list for flasher for models

Resistors
R1   2.7kΩ ¼W
R2   2.7kΩ ¼W

Capacitors
C1   125µF 10V
C2   125µF 10V

Semiconductors
TR1 and TR2   both 2N1304, 2N3053 or BFY51

Miscellaneous
L1 and L2   both 6V 0.1A MES bulbs
2 off MES holders
0.1in matrix board about 6cm × 1cm

Should one lamp light much more brightly than the other, and this continue when the bulbs are interchanged, one transistor may be in poor condition (this is most likely with very cheap or 'untested' devices). If necessary, resistors and capacitors can be checked with a meter. Also make sure connections and supply polarity are in order.

4

Morse Code Practice Oscillator

This oscillator, shown in Fig. 4.1, allows code practice by a single person, or by two or more, as a loudspeaker is incorporated. The circuit, Fig. 4.2, employs two transistors, each driving the other in the familiar multivibrator arrangement. The way in which this operates has been

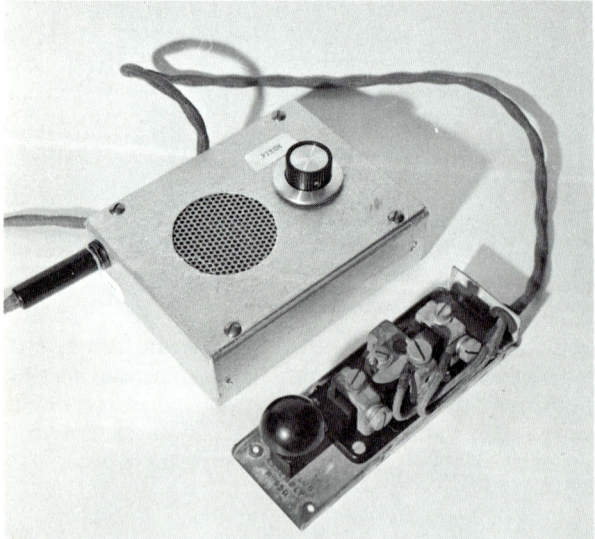

Figure 4.1
Photograph of the Morse code practice oscillator

described in the previous project. The pitch, or frequency of oscillation, is determined largely by the component values, though it is also influenced by the supply voltage, speaker, and individual transistors.

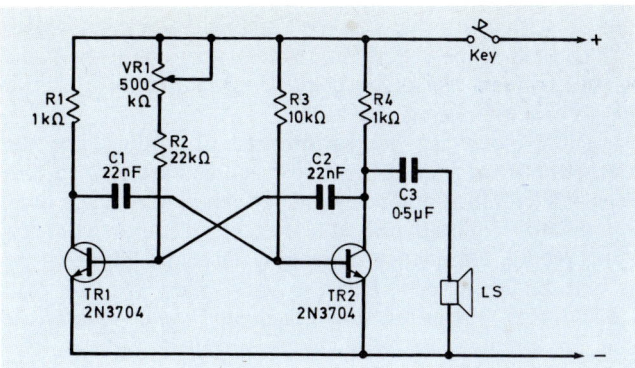

Figure 4.2
Circuit of the oscillator

VR1 is a pitch control, and permits adjustment over a wide range of audio frequencies, to suit the user. This can generally be set to give an audio tone in the 1kHz to 2kHz region, similar to that which would be obtained when receiving code by radio.

The Morse alphabet, and numbers, are as below:

A	·−	N	−·
B	−···	O	−−−
C	−·−·	P	·−−·
D	−··	Q	−−·−
E	·	R	·−·
F	··−·	S	···
G	−−·	T	−
H	····	U	··−
I	··	V	···−
J	·−−−	W	·−−
K	−·−	X	−··−
L	·−··	Y	−·−−
M	−−	Z	−−··
1	·−−−−	6	−····
2	··−−−	7	−−···
3	···−−	8	−−−··
4	····−	9	−−−−·
5	·····	0	−−−−−

Letters should be thought of as combinations of short and long sounds 'dit' and 'dah'. A 'dah' equals three 'dits' in duration. A 'dah' space is left between letters and a slightly longer space between words.

Two persons practising together can send and read alternately. With a group, one may send while the others read. A start has to be made slowly, taking care to form letters correctly. Irregular breaks in a letter may result in F, as example, being read as IN.

A person learning alone can get accustomed to sending, and the sound of the letters. But for reading practice it will be necessary to turn to records, tapes or slow Morse transmissions. To record a practice tape, place the tape recorder microphone at a suitable distance from the loudspeaker, and record a sequence of letters. To avoid filling in by memory, it is best to use random letter squares. These may be read horizontally, backwards, vertically, and diagonally. Two squares are given below, and others can easily be made up, as needed.

```
A S D F G      A Q W S X
Q W E R T      C D R F V
Z X C V B      B G T Y H
P O I U Y      N M J U I
L K J H N      K I O L P
```

When the tape is played back, try to read the message you have just recorded.

Construction

A suitable layout for the Morse code practice oscillator is shown in Fig. 4.3. When fitting TR2, note that the leads are arranged in a different

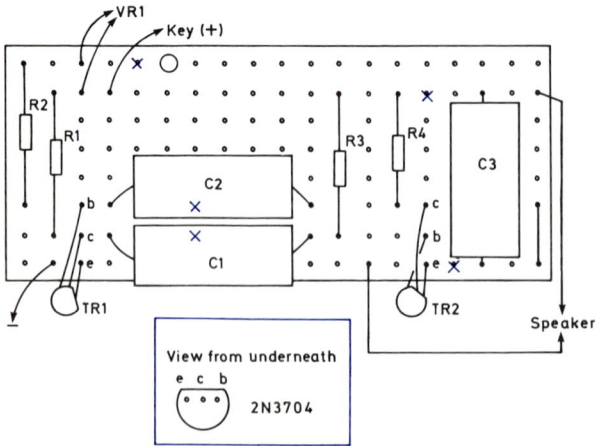

Figure 4.3
Positioning of the components and external leads

position to those for TR1. Also take care to make the foil breaks, cut to avoid shorting C1 and C2, C3 to emitter line and positive line, and in the R2—VR1 conductor.

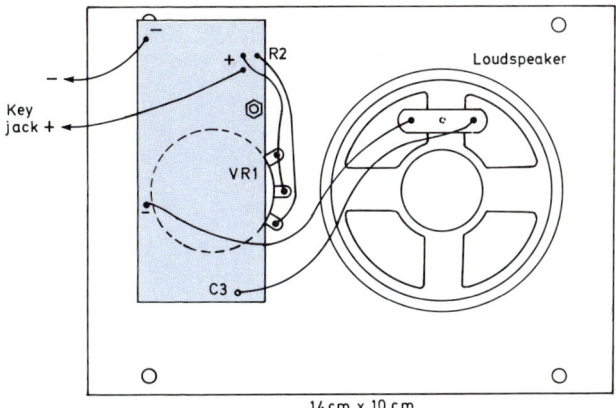

Figure 4.4
Circuit board and speaker fitted to panel

Leads are soldered to the board to allow connecting as shown in Fig. 4.4. The case or box may be metal or insulated material, and the circuits are isolated from the panel and case. A single 6 B.A. bolt through the hole in the board fixes this to the panel, with extra nuts to allow locking it securely, while clear of VR1.

17

The jack for the key plug is fitted to the side of the case. A 3V, 4.5V or 6V supply should be found to be satisfactory, and can be had from a single battery (2-cell or 3-cell) or individual cells in a battery holder.

Table 4.1. Components list for morse code practice oscillator

Resistors
R1   1kΩ (all resistors 5%, ¼W)
R2   22kΩ
R3   10kΩ
R4   1kΩ
VR1   500kΩ or 470kΩ linear potentiometer

Capacitors
C1   22nF
C2   22nF
C3   0.5µF

Transistors
TR1   2N3704
TR2   2N3704

Miscellaneous
Case about 14 × 10 × 5cm, knob, jack socket, plug, cord and key, 0.15in matrix board 7.5cm × 3cm
LS 8Ω to 50Ω 6½cm diameter or similar unit

Any difficulty in obtaining satisfactory results is unlikely. Almost any *npn* transistors to hand may be tried in this circuit. If the pitch proves unsatisfactory with other transistors, a change in the battery voltage may help, or C1 and C2 can be increased in value to lower the pitch, or vice versa. *PNP* transistors are also satisfactory, but the polarity of the battery must then be reversed.

5

Unijunction Transistor Metronome

A metronome is a device used to measure musical time in terms of beats per minute. With its aid, it is possible to maintain regular time through passages of varying difficulty, or to bring a performance up to the required speed. It indicates, by reference to the scale, how rapidly a piece of music should be taken, or how many notes or bars of particular duration should be played in a minute.

The usual clockwork metronome provides a regular beat or 'tick' whose frequency can be changed by altering the position of a weight on a pendulum. In the UJT metronome (Fig. 5.1), a sound similar to that of a musician's clockwork metronome is obtained electronically.

The speed or time may be indicated on the music as a figure showing the number of notes per minute; or it may be expressed as a musical term.

The following list of musical terms is taken from a musician's clockwork metronome, and gives the number of beats per minute:

40	grave	100	allegretto
46	largo	116	allegro
54	adagio	126	vivace
60	andante	144	presto
80	moderato	184	prestissimo

In many instances an *exact* speed is not used, as interpretation will depend to some extent on the musician.

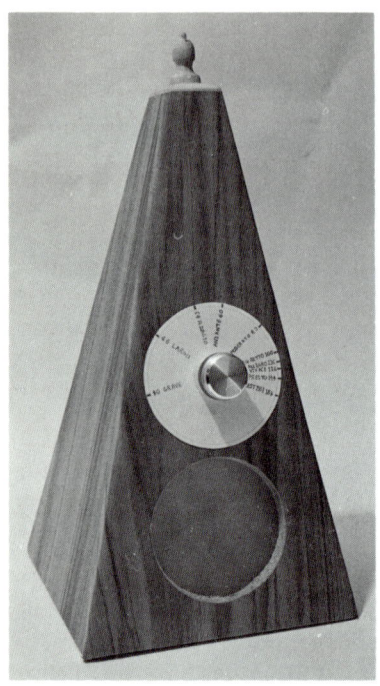

Figure 5.1
The metronome may be fitted into the traditionally shaped box

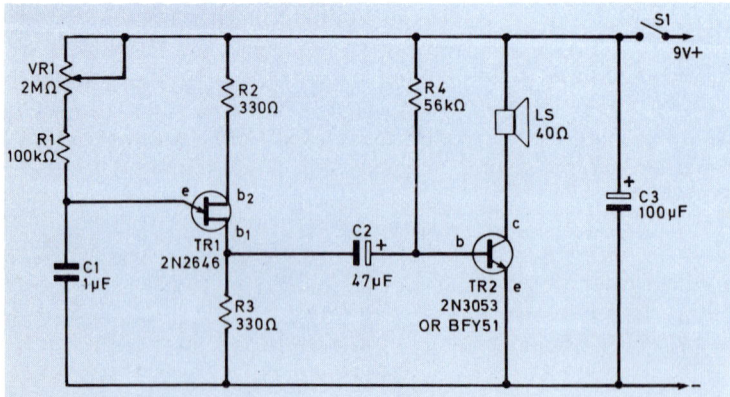

Figure 5.2
UJT timer and amplifier

The circuit

The operation of the timing circuit depends on the unijunction transistor TR1 in Fig. 5.2. Base B1 and base B2 are connections to the opposite ends of the semiconductor material, and a junction of opposite type material is arranged near B2, and forms the emitter E. Supply polarity to B1 and B2 is as shown.

With no bias at the emitter, only an extremely small current flows from B1 to B2, and can be ignored. If the emitter is gradually made more positive, a point is reached where emitter current flows and carriers are available, so that the B1 and B2 resistance falls and current rises rapidly.

In the timing circuit, there is normally no charge in C1, and the emitter is negative. When S1 is closed, C1 begins to charge through the potentiometer VR1 and R1. The emitter voltage thus rises slowly, until the point is reached where the UJT conducts, discharging the capacitor. Each time the UJT conducts, voltage drop in R3 causes a pulse to appear at B1. C1 also begins to re-charge, so that the sequence is repeated. The time taken for C1 to charge depends on the setting of VR1, and grows shorter as its value is reduced. The frequency can thus be adjusted by VR1.

TR2 is an amplifier, receiving base current through R4, and coupled by C2. The beats obtained at B1 of the UJT are able to operate the loudspeaker forming the collector load of TR2.

Construction

The board is 0.15in matrix and measures 5.5 by 4cm. Fig. 5.3 shows the location of components. No foil breaks are required.

Leakage in C1 tends to upset working at low speeds, so a non-electrolytic capacitor is preferred here. The actual component is $1\mu F$, 20%, 250V, but a lower voltage rating is of course in order. With an electrolytic capacitor, a voltage rating rather higher than the metronome supply (9V) is preferred. C2 is not critical, and in any case values such as $47\mu F$ and $50\mu F$ would be completely interchangeable here.

Provide flying leads for VR1 and the speaker, and red and black flexibles for positive and negative. The *on-off* switch S1, in the positive battery lead, is separate from VR1 so that the latter can be left set at any wanted beat.

The unit will operate satisfactorily with speakers of an impedance other than that suggested, though about 35Ω to 75Ω will be best.

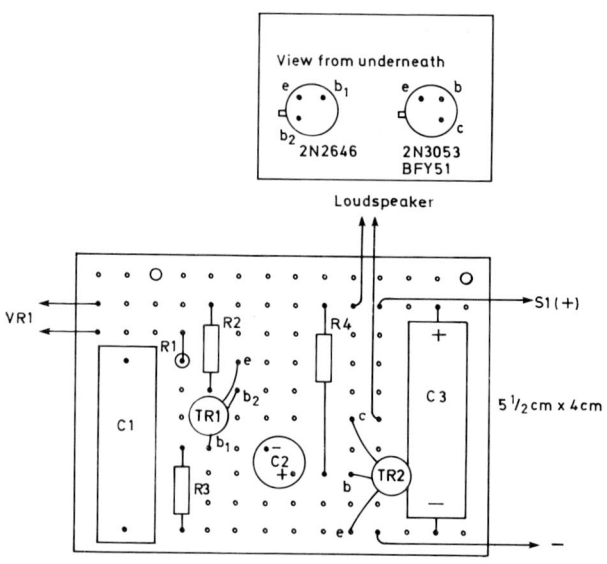

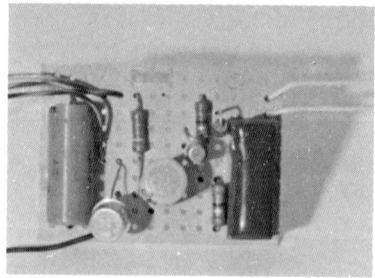

Figure 5.3
The circuit board with components

Fig. 5.4 shows how a typical scale for VR1 is marked out. However, the exact value of C1, and other tolerances, will influence the timing. It is thus better to calibrate the actual scale individually, using a clock or watch with seconds' hand. Adjust VR1 until 40 beats are counted in the minute, and mark this. Proceed in a similar way for the other speeds. The andante setting is of course one beat per second.

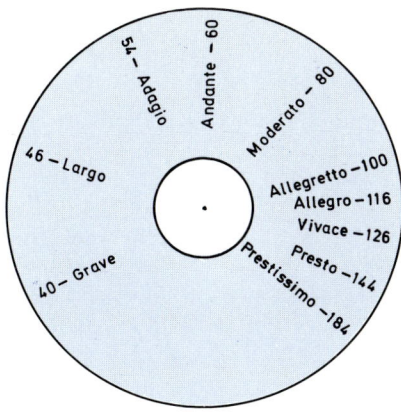

Figure 5.4
Scale for VR1

In view of the difficulty of counting the higher speeds, is it better here to beat time and count each second beat. As example, 72 alternate beats counted for 144.

Constructing the case

The board, battery and speaker could be fitted in any metal or insulated box of suitable type, with VR1 on the top or front. It may however be preferred to construct a case which resembles that of the conventional clockwork metronome.

Dimensions for this are shown in Fig. 5.5. Front, sides, back and bottom can all be 4mm or 6mm plywood. Four pieces are required about 21cm long, 11cm wide sloping down to 2½cm. If the usual type of handicraft power saw and table can be employed, set the saw to 45 degrees, and the corner joints are readily mitred. A block of wood suitably shaped and fitted near the top gives strength here, and the joining surfaces should all fit neatly. Clear off dust, and use a woodworking adhesive. A small square of wood is cemented to the top, and the whole is glasspapered. It may be varnished, or covered with self-adhesive woodgrained material, as used for shelves. All dust must be cleared away, for good adhesion of this material. The back is fixed with small screws run into the block and bottom, so that it can be taken off to replace the battery.

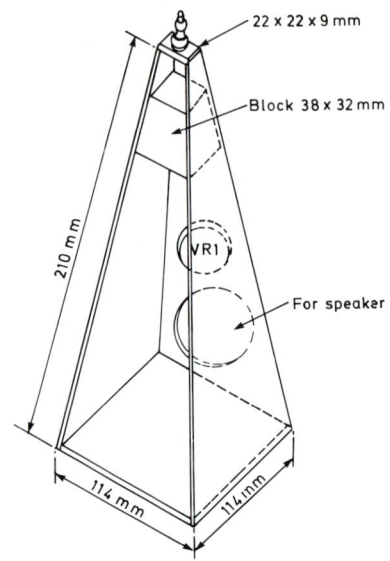

Figure 5.5
Constructional details of the case

Table 5.1. Components list for UJT metronome

Resistors	
R1	100kΩ (all resistors 5%, ¼W)
R2	330Ω
R3	330Ω
R4	56kΩ
VR1	2MΩ linear potentiometer
Capacitors	
C1	1μF paper preferred
C2	47μF 10V
C3	100μF 10V
Transistors	
TR1	2N2646
TR2	2N3053 or BFY51
Miscellaneous	
S1	slide *on-off* switch
LS	approximately 35 to 75Ω, 6½cm unit
	board about 5.5cm x 4cm (0.15in matrix)
	case: as described
	knob, battery clips, etc

The speaker is screwed or cemented in place. Two small screws fix the circuit board, and S1 is fitted to one side of the case. A small top finial and four wooden or rubber feet improve the appearance.

Checking out

Faults causing lack of results are unlikely. If necessary, carefully check connections, and see that nothing has been omitted. If no error can be found, the speaker can be temporarily connected from C2 positive to positive line. If the beat is then audible, the fault must be sought in TR2 or connections here. But if no beat is heard, TR1 is probably not working. Check connections, components, and battery polarity. Leakage in C1 could upset working.

6
Enlarging Exposure Meter

This is an electronic aid to securing suitable exposures when making photographic enlargements. When enlarging, the exposure which is

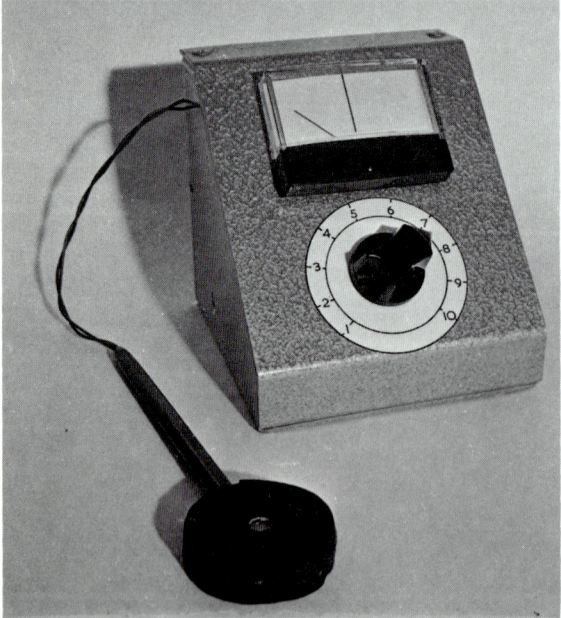

Figure 6.1
The exposure meter and LDR probe

necessary to produce a satisfactory picture when developed may be very brief with thin negatives, a small degree of enlargement, large lens apertures and high illumination. At the other end of the scale will be enlargements from dense negatives, with considerable enlargement, small lens apertures, and limited illumination. Exposures also depend on the paper sensitivity and other factors.

Methods of finding a suitable exposure include obscuring parts of the paper area with an opaque card, which can be moved to allow, as example, exposures of 2, 4, 8, 16 and 32 seconds on the same print; or placing a segmented grey scale over the print. After development, the most suitable exposure time can be seen, and a complete print can be made with this.

The exposure meter shown in Fig. 6.1, once calibrated, will allow a wide range of negative densities and enlargement sizes to be dealt with satisfactorily, without any need for such test strips or trial exposures. This saves both time and materials.

The circuit

The circuit is shown in Fig. 6.2 and uses a light-dependent resistor followed by a single transistor amplifier. When in darkness, or only dimly illuminated, the LDR has a high resistance, so that base current

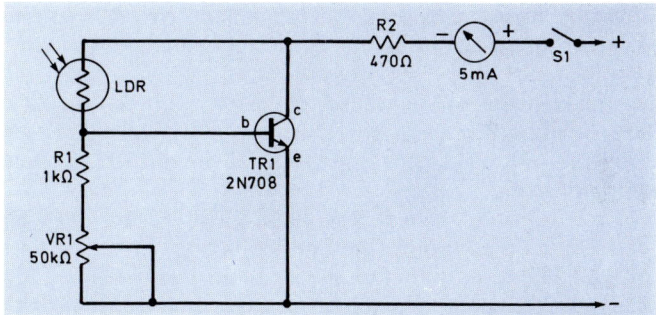

Figure 6.2

Circuit of the exposure meter. VR1 adjusts the sensitivity

to the transistor is small. As light falling on the LDR increases, its resistance falls. Base bias to the transistor is increased and so collector current rises, as shown by the meter.

VR1 is a sensitivity control. Sensitivity is maximum with the whole of VR1 in circuit, falling as the control knob is rotated to provide lower resistance values. R2 is merely to limit current, as the meter may easily be driven beyond its full-scale reading.

The LDR is fitted to a handle with flexible leads, so that it can be moved about on the enlarger baseboard (Fig. 6.3). It is also covered by

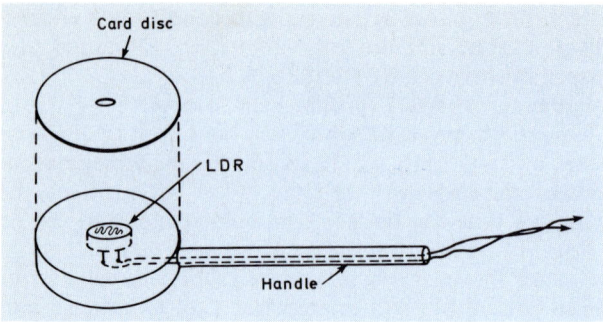

Figure 6.3
LDR arranged to place on the enlarger baseboard

an opaque card having a hole, so that small sections of the image can be checked for brightness.

Construction

This uses matrix board measuring about 7.5cm by 2.2cm. The components are arranged as in Fig. 6.4. A smaller board could be used, but this allows plenty of room.

Only one break in the foil is needed, under R1. This can be made with a cutter, or with a few turns of a sharp drill. More breaks arise in later units, and it is essential that fragments of foil are not left to touch adjacent conductor strips.

Two holes are drilled to take 8 B.A. bolts about 12mm long. By using two extra nuts on each bolt, the finished board can be spaced a little clear of the metal case, so that no short circuits can arise.

Flying leads are provided for external circuits. It is convenient to use colour coding for such purposes, red for battery positive circuit, and black for negative, as a minimum.

A check should be made that all soldered joints are sound, and that no unwanted contact or circuit is provided between adjacent conductors.

Case

As with many other projects in this book, there is considerable latitude in the type and size of case. Cases and boxes can be obtained in a wide variety of types, both plastic and metal.

A sloping front case is probably most suitable here (Fig. 6.1), as it allows the meter to be seen easily. An opening for the meter can be cut

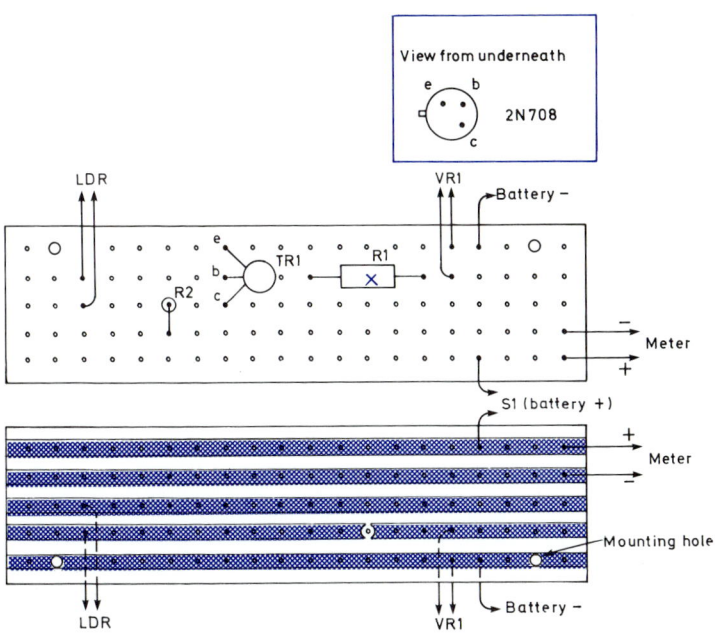

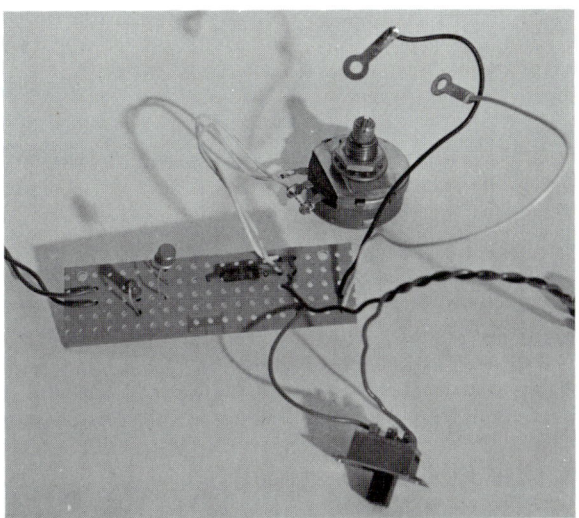

Figure 6.4
Board and components

with an adjustable tank cutter, or with a screw-up punch; or it can be made by drilling a ring of small holes, closely together, and smoothing off with a half-round file.

The unit can be operated from 4.5V to 9V, the latter providing the greater sensitivity. As battery current is small and intermittent, a miniature 9V battery is adequate.

Method of indication

Readings have to be taken with only the enlarger light on, and so it is convenient to use the circuit in such a way that VR1 is adjusted so that the meter pointer always comes to rest in its centre position. This should be marked as a reminder and for easy observation in poor light.

It will be found that the instrument can be used in several different ways.

(1) With a given grade of paper and development, note the setting of VR1 which results in full density just being obtained on any thin part of the negative image. Lens aperture can then be adjusted to bring illumination to this same level with equivalent parts of other negatives. That is, a standard brightness level, as shown by the meter, is adopted. Then exposures with the same timing will provide similar results. If there is considerable difference in degrees of enlargement and negative density, two or more such standard brightness levels, each with its own exposure time, will need to be noted, for different settings of VR1.

(2) A list of exposure times can be drawn up for various settings of VR1. Once this is done, it is possible to read off suitable times for full density of thin parts of any negative.

When first bringing the instrument into use, it should be noted that the exposure-aperture relationship which exists when taking photographs will often not hold with the enlarger. This depends on the type of lighting (diffused or condenser) and the extent to which enlargement is being obtained. Normally, the next stop down would double exposure. But if condenser illumination directs a sharp cone of rays through the lens aperture at some degrees of enlargement, stopping down will have less effect on illumination. It will also generally be found that illumination is weaker towards the edges of the image.

It is thus necessary to arrive at an initial calibration by making a trial exposure in one of the ways described earlier. But once this has been done correctly, it need not be repeated as readings will subsequently be found by placing the LDR on the enlarger baseboard.

Other points

A 5mA meter has been used in this project, but other more sensitive movements can be adopted if to hand. With a 1mA instrument, R2 can

Table 6.1. Components list for enlarging exposure meter

Resistors
R1   1kΩ (all resistors 5%, ¼W)
R2   470Ω
VR1   50kΩ (47kΩ suitable) linear potentiometer

Transistor
TR1   2N708

Miscellaneous
LDR   light dependent resistor (ORP12)
Meter   5mA (38mm)
S1   slide switch
            board about 7.5 × 2.5cm
            sloping front case 10 × 10 × 10cm
            battery connectors, knob, etc

be increased to about 2.2kΩ. The actual value is not important, as this component is merely used to limit maximum current through the meter, should VR1 be adjusted carelessly, with the LDR strongly lit.

Various other transistors will operate in this circuit. That shown is the *npn* type. If a *pnp* transistor were used, battery and meter polarity would have to be reversed.

7
`Freeze´ Indicator

This unit indicates when the temperature has fallen to a figure which may be dangerous or harmful. The temperature at which the warning is given can be adjusted between wide limits, and the temperature-sensitive element can be at any reasonable distance from the indicator unit.

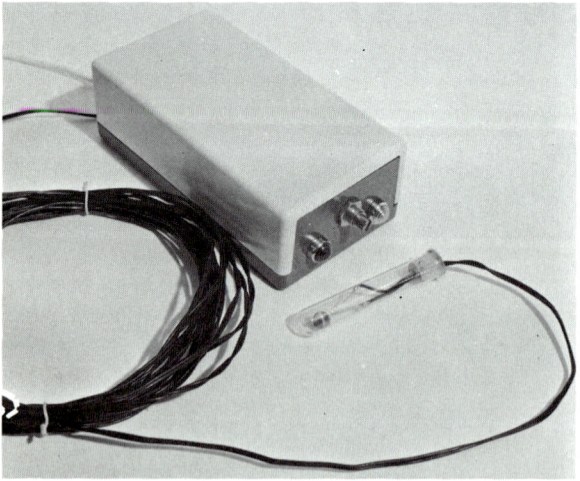

Figure 7.1
The portable 'freeze' probe and indicator unit

It can be used in a greenhouse or garage, to show when the temperature is likely to reach freezing point, or in similar circumstances where plants or livestock may be harmed, or when attention should be called to the need for heating or other precautions. A drop in temperature taking place outside may easily be overlooked, and the indicator will also show if a glasshouse or other building heating system has failed, so that conditions are not being maintained at a safe level.

The circuit

Fig. 7.2 shows the circuit, and operation is controlled by the thermistor VA1040. The resistance of this item depends largely on its temperature.

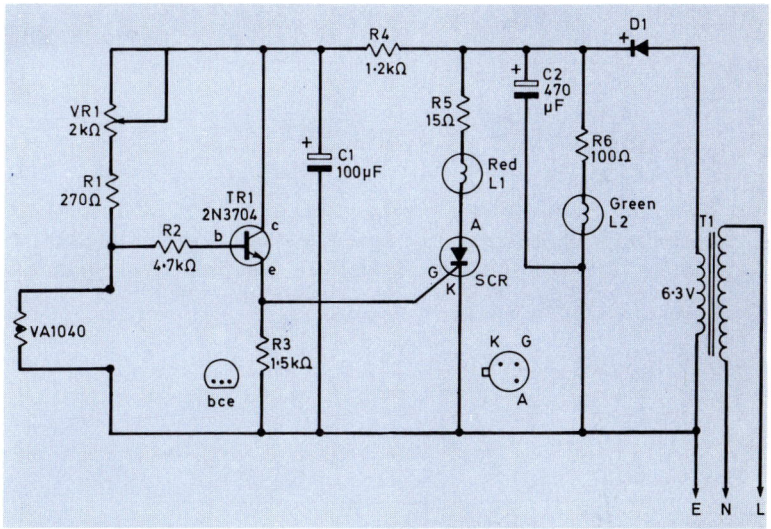

Figure 7.2
Circuit of the temperature controlled or 'freeze' indicator

Typically, it will be about 250Ω at 15°C (60°F), rising to about 450Ω at 4°C (40°F), and 1200Ω at −12°C (10°F). Base bias conditions for TR1 depend on the thermistor R1 and the setting of VR1. As the thermistor, on the one hand, and R1 with VR1, on the other, form a potential divider, VR1 can be set so that TR1 begins to pass emitter current when the thermistor value has risen to any wanted resistance. A drop in temperature thus moves TR1 base positive, until the voltage drop across R3 triggers the silicon controlled rectifier into conduction.

C1, with R4, prevents a surge when switching on, which might trigger the SCR. Operating current is derived from the 6.3V transformer T1, with half-wave rectification by D1. About 7.5V will be obtained across the reservoir capacitor C2.

L2 is a green indicator lamp, to show that the unit is plugged into the mains supply and operating. R6 reduces the filament voltage for

33

long life, and 100Ω will be correct for a 6V, 0.1A lamp, with a 6.3V transformer.

L1 is a red lamp, and comes on only when the temperature has fallen to an extent decided upon. A 6V, 0.1A bulb is also used here, and requires a series resistor R5 of about 15Ω.

The SCR is a small 50V, 1A type, with lead-outs as shown. It does not conduct until sufficient gate current is available, and then remains in the conducting state until anode voltage is removed.

Operating current is obtained from the mains, by means of the transformer T1, so that the unit can be left on permanently. It can be hung on a wall or placed in a conspicuous position indoors, with an extension lead running to the temperature dependent resistor.

Construction

Fig. 7.3 shows the positions of components on a 7.5 × 2.2cm board of 0.15m matrix. Foil breaks will be noted between R1 and TR1 emitter, R2 and R4, the SCR cathode K and D1 negative, between gate G

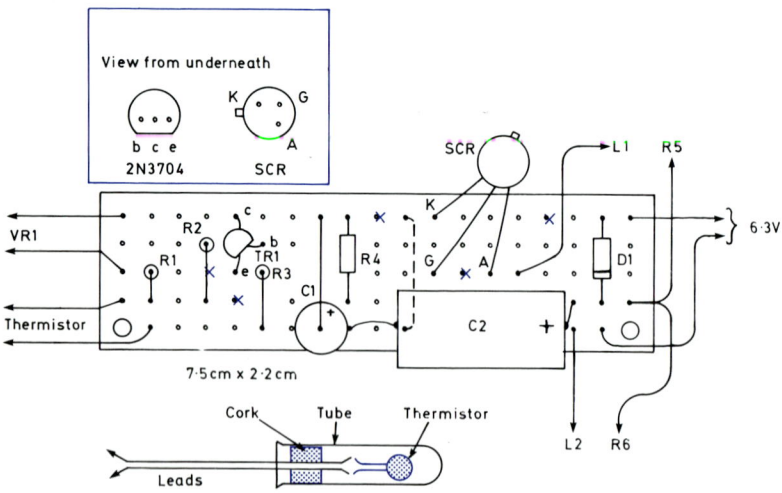

Figure 7.3

Positions of the components on the board, and the thermistor sealed into the tube

and anode A of the SCR, and between R4 and SCR cathode. An insulated lead is soldered on as shown to complete the circuit from the SCR cathode K to the negative line near C2 negative.

There are several external circuits to be completed by the leads provided. R1 and TR1 collector circuit are taken to the sensitivity potentiometer VR1, mounted on the case. R1 and the negative line run to the thermistor, which is arranged as described later.

At the other end of the board, connections run from the SCR anode A and D1 positive to lamp L1 and R5. R5 is soldered directly to one lampholder tag. D1 positive and the negative line are similarly connected to the lampholder for L2, with R6 at the holder.

Finally, a lead from the negative line goes to the 6.3V secondary of the transformer, and also to earth E. D1 negative is connected to the remaining secondary tag.

Thermistor

This needs to be protected from wet or damp. The easiest way to do this is to solder the extension leads to the thermistor leads, and insert the thermistor in a test tube, as in Fig. 7.3. A cork is then pushed in and sealed with wax or similar means.

Tie or hang the tube in a suitable position in the greenhouse (or where required) and run the leads by the most convenient route to the indicator.

Case

The type of box or case will not be important, though a metal box should be earthed by the earth conductor of the 3-core mains cord. No mains *on-off* switch is included.

Fig. 7.4 shows the transformer, circuit board, and other items fitted to an insulated box 15 × 8 × 5cm. Connections to the board are shown in Figs. 7.3 and 7.4.

The spindle of VR1 is slotted for screwdriver adjustment, as generally this will be set in the way described later, and will be left in this position.

Adjustment

The green lamp L2 should glow when the unit power plug is inserted. The red lamp will light when VR1 is turned for minimum resistance, but not when the whole element is in circuit. To extinguish L1 it is necessary to switch off at the mains outlet, then on again, for adjustment of VR1. Set VR1 so that L1 fails to light at a moderate room temperature. The temperature sensing element can then be placed, with a thermometer, in the part of a refrigerator where a temperature of about 4°C (40°F) will be obtained. After a few minutes' delay, according to temperatures, L1 should light.

A fairly accurate adjustment is possible provided the thermometer used to check temperatures and the thermistor tube are close together,

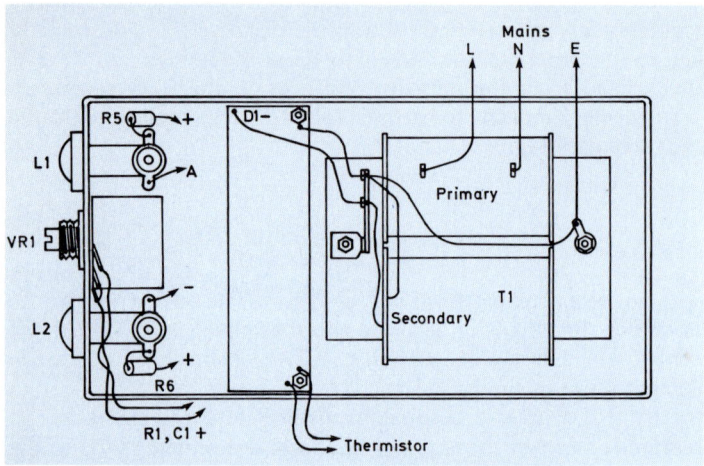

Figure 7.4

Assembly in 15 × 8 × 5 cm insulated Verobox

and enough time is allowed for the thermistor to take up the temperature of its surroundings. If the SCR is triggered before the temperature has fallen to a low enough level, carefully re-adjust VR1 so as to increase the resistance in circuit. When first adjusting VR1, a rough approximation can be found by turning it one way or the other, to find the position which causes L1 to light.

Table 7.1. Components list for 'freeze' indicator

Resistors
R1 270Ω, ¼W
R2 4.7kΩ, ¼W
R3 1.5kΩ, ¼W
R4 1.2kΩ, ¼W
R5 15Ω, ½W
R6 100Ω, ½W
VR1 2kΩ linear potentiometer
Thermistor VA1040

Semiconductors
TR1 2N3704
SCR 50V, 1A or similar silicon controlled rectifier
D1 50V, 1A silicon diode

Miscellaneous
T1 6.3V transformer, ½A rating, or similar
L1 6V, 0.1A bulb and holder (red)
L2 6V, 0.1A bulb and holder (green)
C1 100μF, 12V
C2 470μF, 12V
tube, knob, etc
vero box 15 × 8 × 5cm
3-core mains cord

8
Adjustable Power Supply Unit

This power supply unit runs from a.c. mains, and will provide any required output from 0V to 12V, at a current of up to about 1A. Within these power limits it may be used to run a receiver, amplifier, tape recorder, or similar equipment. It will also provide power for

Figure 8.1

The adjustable power supply unit

experimental circuits, or for various projects such as described in this book.

It is also suitable for other purposes, which include the slow charging of accumulators or rechargeable cells such as may be used with models, or for 'rejuvenating' ordinary dry cells. It will also provide current for electroplating, or for small model motors, or for the illumination of models.

The power-supply unit (PSU) has automatic overload protection, to avoid damage to its components should an accidental short circuit arise in the output when connecting or operating a model.

The circuit

The circuit is shown in Fig. 8.2. Current is obtained from the transformer T1, which reduces the voltage, and also isolates the low voltage and output part of the equipment from the mains. S1 is the main *on-off* switch, and the neon panel indicator shows when the PSU is on.

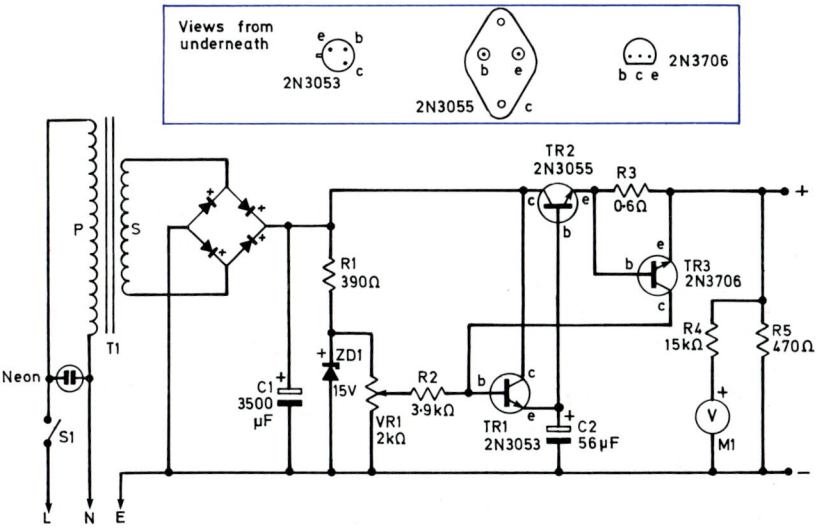

Figure 8.2

Low voltage a.c. from the transformer is rectified and passed as d.c. by TR2 which is controlled by TR1. TR3 provides short-circuit protection

The secondary S and rectifiers provide d.c. for the reservoir capacitor C1. R1 supplies current to the Zener diode ZD1, so that a stabilised potential is obtained for the voltage control potentiometer VR1. The voltage across VR1 thus remains almost constant, despite changes in current drawn from the PSU.

TR1 is a current amplifier to supply the base of TR2. The base voltage of TR1 is adjusted by VR1, and its emitter closely follows this, thereby allowing adjustment of TR2 base potential. TR2 is a power transistor passing the PSU output circuit current, in which the

39

emitter also closely follows the base voltage. Adjustment of VR1 thus allows TR2 emitter to be adjusted from about zero to somewhat over 12V, the emitter potential of TR2 being substantially independent of current drawn. C2 removes residual hum from TR2 base, for electronic smoothing of output.

The actual output voltage is shown by a voltmeter, or by the 1mA instrument with series resistor R4. R5 draws a steady current to improve performance at very low current levels.

TR3 provides automatic current limiting under short circuit conditions. Current flows through R3, and normally TR3 is not conducting. If a short circuit (or heavy load) arises, a high voltage drop across R3 causes TR3 base to be positive so that TR3 conducts, meanwhile the excess load having moved TR3 emitter negative. Collector current through R2 causes a substantial voltage drop in this resistor, and TR1 base is moved negative. TR1 emitter and TR2 base follow, so that TR2 emitter moves negative, limiting short circuit current to a safe level. When the short is removed, operation returns to normal.

Construction

The circuit is best assembled wholly on the panel, as in Fig. 8.3. It can then be tested readily without its case.

When wiring up the primary circuit of T1, clamp the 3-core cord securely. The transformer, panel, and negative line of the output are earthed. Current should be drawn from a 3-pin plug, correctly connected, and with a 3A or other small rating fuse. The usual neon indicator will have its series resistor incorporated.

The heat sink for TR2 is flanged aluminium, about 10cm x 5cm. Drill the four holes for TR2, and remove burr, so that the transistor rests flat on the sink. Emitter and base pins pass through clearance holes. The sink is fixed to the panel by two 6 B.A. bolts, but is electrically isolated from the panel itself. This is done by placing a 2mm thick strip of paxolin or similar material about 10cm x 1.5cm between sink flange and panel, and using bushes or insulated washers under the nuts.

Wiring can then be completed as in Fig. 8.3. Note that tags 4 and 9 of the long strip are used to mount it to the sink, so are common to TR2 collector. Fig. 8.4 shows the strip. A 50V, 1A bridge rectifier may be used instead of the four diodes. This will have four tags or leads. Take the a.c. input leads to 6 and 7 on the strip, with positive to 4 and negative to 1. The polarity marking for 1N4001 diodes is as in Fig. 8.4.

TR3 and its associated components mount on the small strip. This is secured by one of the meter fixing nuts.

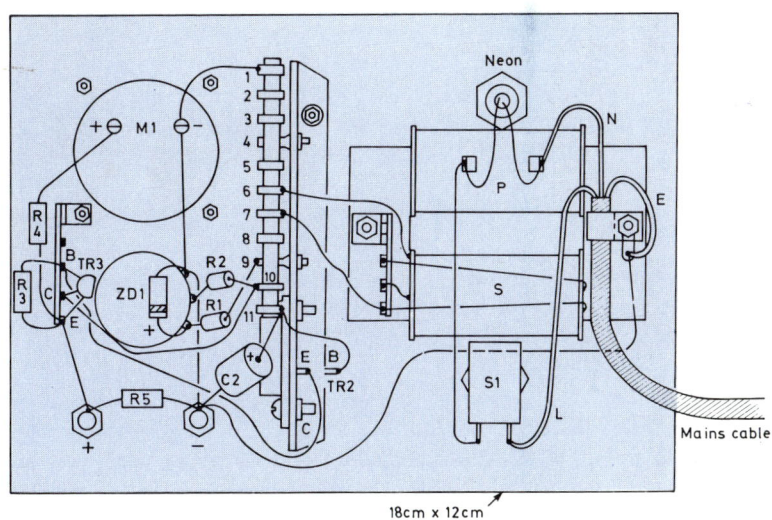

Figure 8.3
Assembly on panel

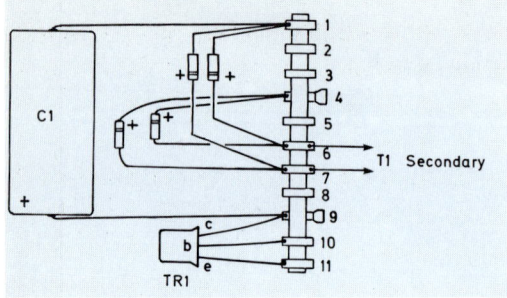

Figure 8.4

Rectifier diodes, C1 and TR1 assembled on the strip

Meter

A voltmeter reading 0–12V or, ideally 0–15V may be used, and can be purchased ready scaled and with an internal series resistor. If such an insturment is used, R4 should be omitted.

With a 1mA meter, R4 of 15kΩ (1%) allows this to indicate 0–15V. For 0–12V, R4 would be 12kΩ 1%. In each case the additional resistance of the meter itself should be deducted, for maximum accuracy. As example, using a 100Ω 1mA instrument for 0–15V, R4 ought to be 14,900Ω. Such a resistor value is not readily obtained, but it can be closely approached by using 15kΩ and 2.2MΩ resistors in parallel. (An alternative is to select a slightly low 15kΩ resistor by means of an accurate test.)

Other meters can be used by changing R4 to suit. The total series resistance required is V/I. As example, for 15V with a 5mA meter, 15/0.005 = 3,000Ω or 3kΩ.

With other than a ready scaled meter, it will be necessary to remove the front and draw up a suitable scale on thin card. Take care not to disturb the pointer or internal parts.

Transformer

Low-voltage transformers often have optional taps. Unwanted taps are left unused, but flying leads need to be taped, or soldered to unused tags, so that they cannot touch each other or other conductors.

A secondary rated at 16V to 17V is required here. After rectification, this will provide around 23V across C1. R1 thus has to drop about 8V. Should the secondary voltage be rather different, R1 can have 50Ω resistance for each 1V to be dropped, and this will keep power dissipated in the Zener diode well within its rating.

Selecting R3

The exact limiting point will depend somewhat on individual samples of TR3, but is not too critical. A short or overload is likely to be cleared quickly so no damage to the PSU components is probable even if current slightly exceeds 1A.

However, it is easy to adjust R3 so that a meter indicates that short circuit current is near to 1A maximum. Lowering the value of R3 increases the short circuit current level. Generally, a resistor of 0.56Ω, or two 1.2Ω resistors in parallel, should be satisfactory. If only up to about 500mA current will be needed for electronic equipment, R3 can be one 1.2Ω resistor, or two 2.7Ω or similar resistors in parallel.

Testing

If a fault occurs, check that base and emitter pins of TR2 are not in contact with the sink, and that the sink is isolated from the panel and negative line. A d.c. voltmeter should show about 23V across C1. The base and emitter of TR1 and TR2 should follow the rotation of VR1.

Table 8.1. Components list for the adjustable PSU

Resistors		
R1	390Ω, ½W	
R2	3.9kΩ, ¼W	
R3	0.6 or 0.56Ω (see text)	
R4	15kΩ 1%, ¼W (see text)	
R5	470Ω, 1W	
VR1	2kΩ, ½W linear potentiometer	
Capacitors		
C1	3500μF, 25V	
C2	56μF, 25V	
Semiconductors		
Rectifiers	4 × 1N4001	
ZD1	15V, 400mW	
TR1	2N3053 or BFY51	
TR2	2N3055	
TR3	2N3706	
Miscellaneous		
M1	0−12V or 0−15V voltmeter, or 1mA meter (see text)	
T1	mains transformer, 16V, 1A secondary	
	case with panel about 19 × 14 × 9cm	
	heat sink (10cm × 5cm), knob, mains toggle switch, indicating neon fitting, output sockets, tag strips, etc	

9

Light Actuated Receiver and Lamp Unit

The light actuated receiver is primarily intended for use with the lamp unit described later, but it can also be controlled by other light sources. With a model train, for example, only a short distance need separate a small lamp and the light actuated receiver, for automatic off switching when an engine moves into the light path. In fact any suitable light source can be utilised, even including daylight, where the unit can automatically switch on artificial lighting when darkness falls.

The maximum range over which the equipment can be operated depends on the intensity of the light source and the extent (if any) to which this is focused into a narrow beam, and upon the general level of illumination present. Most sensitive operation is achieved in subdued light, or darkness.

The circuit

The basis of the light actuated receiver (Fig. 9.1) is the light dependent resistor, LDR in Fig. 9.2. When the LDR is illuminated, its resistance is only some hundreds of ohms, but this rises to many thousands of ohms in darkness or with much reduced illumination.

Base bias for TR1 is obtained by the potential divider formed by VR1 with R1, and the LDR. With the LDR illuminated, and thus of low resistance, TR1 base is negative, and both emitter and collector currents are small. When light fails to reach the LDR, its resistance rises, so that the base of TR1 moves positive. The illumination level at which this becomes significant can be adjusted by VR1, which is the sensitivity control.

Emitter current through R3 causes a voltage drop in R3 and so moves the base of TR2 positive. This causes TR2 collector current to rise, energising the relay.

Figure 9.1
Light actuated receiver

The relay has contacts A and B. When it is energised, these contacts close. When contacts A close, current is provided for the indicator lamp, from the 5V tap on the transformer. This lamp is useful in setting up the unit in working condition. Sockets are provided in parallel with the lamp, enabling an extension circuit to be controlled as well. Leads may run to a bell, buzzer, or other warning or indicating device.

If interruption of light reaching the LDR is momentary, the circuit reverts to its original condition, and the relay contacts open. This is convenient in some circumstances, as when using the equipment to show that a person has passed into a shop or a separate and perhaps distant salesroom. In other circumstances (as when used for a burglar alarm) S1 can be closed, to make use of the 'latch on' facility. Relay contacts B close at the same time as contacts A, and so current flows through R4, contacts B, S1 and the relay coil, holding the relay on. It stays in this condition until S1 is opened.

Diode D1 is a safety device, to avoid the possible damage to TR2 which might arise from back e.m.f. generated in the relay winding.

Transformer T1 is a bell transformer, with 5V and 8V taps. The operating voltage is not very important, but it is convenient to use the 5V tap for the lamp indicator and extension circuit, and 8V for the rectified supply obtained from D2, for the reservoir capacitor C1.

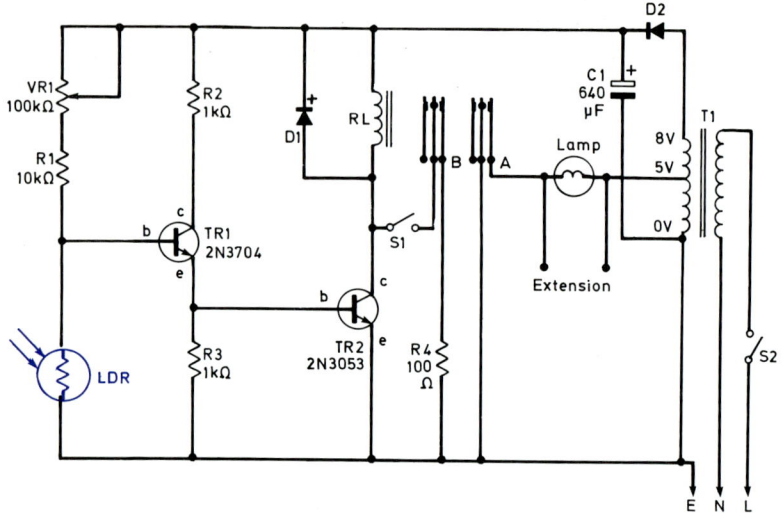

Figure 9.2

Circuit of the receiver unit. S1 provides latch-on of the relay

For stopping toy trains or other models, battery running is practical. T1, C1, S2 and D2, with associated wiring, can then be omitted, and a 9V battery (or similar supply) may be employed instead.

Construction

Fig. 9.3 shows the matrix board which should measure approximately 10cm × 4.5cm. A small angle girder, 12cm long, is used to mount the board to the flanged sides of the case. This part can be bent from a piece of 20swg aluminium or other thin metal, about 12cm × 2cm in size, by gripping it in a vice. Or a similar piece can be cut from a ready-shaped 'universal chassis' flanged member. A tag or lead joins the metal to the second foil, to provide a common or negative return from the case to the negative line of the board.

The LDR is mounted by inserting two pins, and soldering short, stout leads to these on top of the board. This permits easy adjustment of the LDR height and position, so that it will match up with a 1cm hole in the case.

Some free space has been left around the relay, as the size and shape of this component can vary somewhat. A small 170Ω relay was used here, but any relay with a coil of about 100Ω to 250Ω resistance, and which closes reliably with the voltage available, will be suitable. R4 has to pass sufficient current to hold the relay closed, once it has operated, and the value shown should normally be satisfactory.

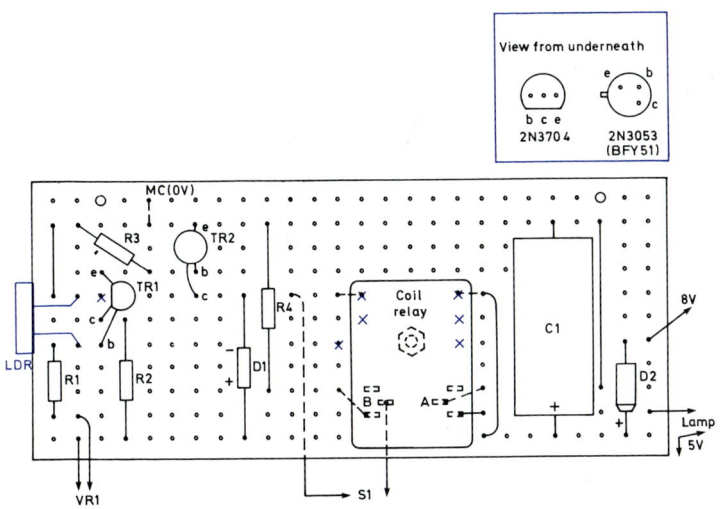

Figure 9.3

The placing of the components on the board

The relay has double-pole change over contacts. Connections are to the contacts which close when the relay winding receives current. (Should it be wished to switch a device or model *off* when light is interrupted, use the alternative contacts, which open when the relay is energised.)

A foil break is required under TR1, and around the relay fixing screw so that adjacent conductors are not shorted together. Clearance holes are drilled for the coil pins, and a slot is cut for the contact tags. The coil is connected from TR2 collector to positive line. Flying leads can be left for other connections, or can be soldered on later, as required.

Case and fitting

The unit is built in an open case formed from a single 32cm × 5cm 'universal chassis' member. A 90° section is cut out of each flange,

10cm from the member ends. Right-angle bends at these points then produce the 12 × 10 × 5cm open box shown in Fig. 9.4.

A cover is made from aluminium or other thin metal, measuring 25 × 12cm. This is bent into the required shape (12 × 10 × 5cm) by

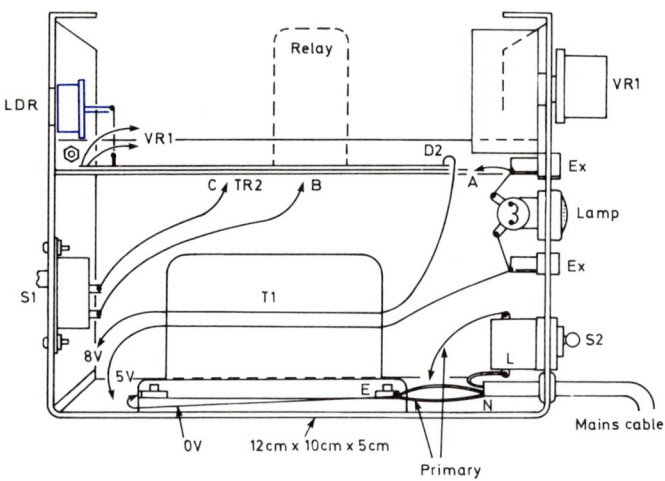

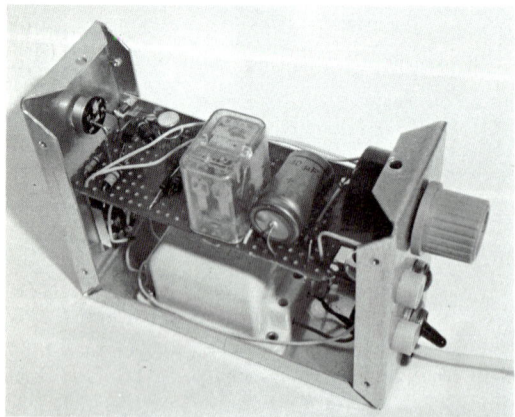

Figure 9.4
The circuit board and other components fitted into the case

gripping it in a vice, or between boards or metal angle clamped level with the required bending line. It is later secured by means of self tapping screws, carefully positioned so that they do not touch any internal connections or parts.

48

In Fig. 9.4 the left-hand side of the case carries S1, and has the opening for the LDR. The right-hand end or side has VR1, lampholder, pair of extension circuit sockets, mains switch S2, and a grommet for the mains cord.

When the board is fixed in the way described, connections can be made to it from S1, VR1, T1 and the lamp circuit (Fig. 9.3).

Use a 3-core flexible cord for the mains lead, correctly connected to a 3-pin plug, so that the metal case is earthed. The plug should have a 3A or other low rating fuse inserted. The earth lead is grounded at a tag placed on one of the bolts holding the mains transformer. Mains neutral goes to the transformer primary, and the live mains conductor to the switch S2.

Adjustment

Initially have S1 open, and face the LDR towards the light. Set VR1 so that the indicator lamp is switched on when the hand is positioned to obscure the light falling on the LDR.

As the general level of illumination will also cause a change in the resistance of the LDR, it is necessary to set VR1 to suit conditions.

Table 9.1. Components list for light actuated receiver

Resistors		
R1	10kΩ, ¼W	
R2	1kΩ, ¼W	
R3	1kΩ, ¼W	
R4	100Ω, ½W (resistors 5%)	
VR1	100kΩ linear potentiometer	
Capacitor		
C1	640μF, 20V	
Semiconductors		
TR1	2N3704	
TR2	BFY51 or 2N3053	
LDR	ORP12 or similar	
D1	50V 1A silicon diode rectifier	
D2	50V, 1A silicon diode rectifier	
Miscellaneous		
Relay	170Ω 2-pole 2-way	
Bulb	6V, 0.1A with holder	
S1	*on-off* slide switch	
S2	*on-off* mains voltage switch	
T1	bell transformer, typically 0–5–8V, 1A	
	0.15in matrix board approximately 10 × 4.5cm	
	materials for case etc as described, sockets etc	

With the lamp unit described next, no difficulty should arise in setting up the equipment to operate reliably at a distance of some 10m to 12m or so, and considerably greater ranges can be obtained.

It is necessary that any indicating lamp etc. connected to the extension warning circuit should not exceed the ratings of the relay contacts or T1.

Mains voltages cannot be switched directly by means of the small relay incorporated.

The unit can be placed flat near the track for model trains, with an illuminating lamp fitted opposite.

Construction of the lamp unit

The lamp unit is shown with the light actuated receiver in Fig. 9.5. For long periods of use this unit needs to be mains powered, so a 6.3V bulb is used with a 6.3V transformer. For short periods of running, with a model, a battery may be employed instead, or it may be possible to derive current from a train or other power supply.

Figure 9.5

The lamp unit and receiver illustrated together

The mains operated unit is constructed as in Fig. 9.6 and is wholly enclosed. The case is made from a 30 × 5cm flanged universal chassis runner, 90 degree sections being snipped out of each flange 10cm from the end, so that the part can be bent into a 10 × 10 × 5cm open box. A cover is made from a flat 25 × 10cm sheet, bent to fit. This is most easily done by clamping the metal between boards or angle steel level with the bending line, as already described. The shaped cover is fixed in place with self tapping screws.

For working over a short distance, no lens is necessary. But this is essential for greater distances, as it concentrates the light into a beam. A lens taken from a hand lamp may be used. It is fitted in a small metal canister with screwed top. This was done by cutting a hole a little smaller than the lens in the can lid, so that it projects as in Fig. 9.6. Small sections of the lid perimeter were turned over the rim of the lens, to help hold it. Similar pieces turned inwards around the canister top allow the lens to be gripped securely when the lid is screwed on.

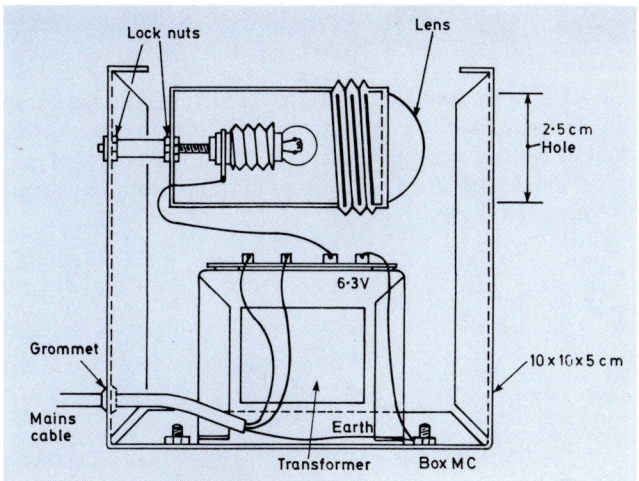

Figure 9.6

Lamp with mains transformer. The lens helps focus the beam

A 4cm long 6 B.A. bolt is passed through the lampholder, remembering to keep the insulated washers in place, and the bulb is arranged as shown. For working at maximum distance, adjust the spacing between bulb and lens to that which gives the most efficient beam, and lock the holder in this position. The lens should be of short focal length (have considerable curvature) but need not of course be of high optical quality.

For a 6.3V transformer, the bulb can be 6.3V, 0.5A, 0.3A or 0.15A, depending on the level of illumination required. A hole about 2.5cm in diameter is punched in the case level with the lens.

Table 9.2. Components list for lamp unit

Mains transformer, 6.3V, ½A or 6.3V, 1A secondary
Bulb 6.3V, 0.15A to 0.5A, and holder
Universal chassis flanged side, 30 × 5cm
Sheet metal for cover, 25 × 10cm
3-core mains cord
Lens and materials for fitting as described

It is possible to fit an infra-red filter over the aperture, and the beam is then virtually invisible, but this considerably reduces the sensitivity, and is not required for the purposes described. With the bulb and lens correctly positioned, and the latter set back from the opening as shown, most light will emerge as a beam.

10

Home Intercom

These units will provide communication between two rooms of the house or two other points. Typical uses include communication between a children's play room and kitchen, or kitchen and living room, or from a bedroom to downstairs room. Alternatively, communication between house and an out-door toolshed or workshop can prove to be very convenient.

Figure 10.1
The master unit and extension

Children may use the communication units merely for games, with a temporary extension line. The units may also be employed as a 'baby cry' listening device.

With only a very small increase in complication of circuitry it is possible to avoid the need for a battery (which is employed in some commercially produced systems) in the extension units and to allow calling both ways, as well as speech, with a 2-wire extension line. So this is the method employed with these units, which run from a single 9V battery.

53

The circuit

Fig. 10.2 is the circuit of the larger, main unit. It is quite possible to use other amplifiers, though the 3-transistor circuit, with a simple class A output stage, is adequate for this type of equipment.

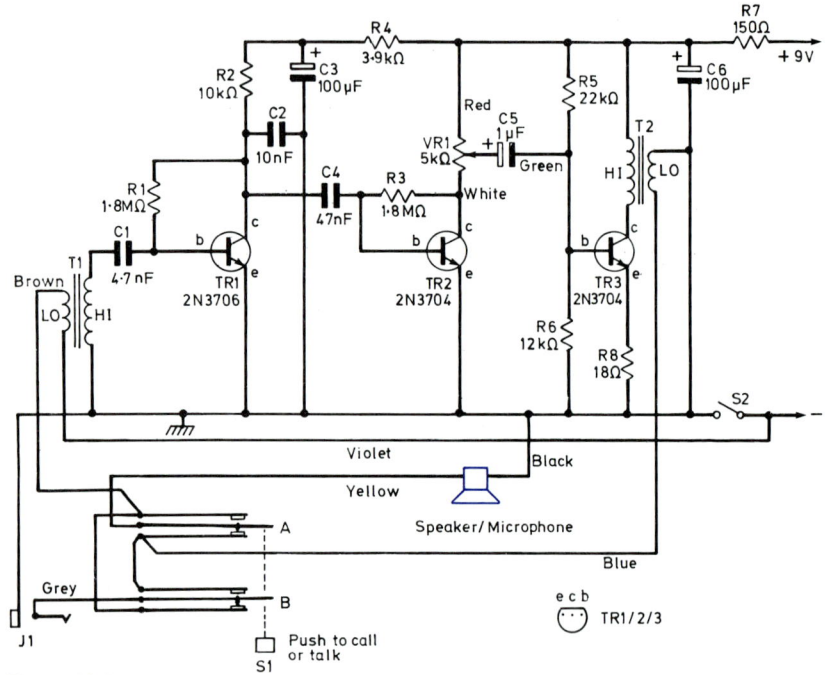

Figure 10.2
Circuit of the master unit

TR1 is the first audio stage, with audio input via C1 to the base B. R1 supplies base current and R2 is the collector load. C4 couples audio signals to the next amplifier TR2. C2 is a bypass capacitor, to help maintain stability. C3 and R4 also decouple the supply to the first stage TR1.

VR1 is the audio gain or volume control, with C5 coupling to the last stage. Operating conditions here are set by the base resistors R5 and R6, and emitter resistor R8. TR3 passes about 20 to 25mA, so giving reasonable loudspeaker volume for this kind of equipment.

T1 is the input coupling transformer, and T2 a similar transformer used for output coupling. Both components are about 12:1 ratio, with 2Ω or 3Ω speaker units operating either as microphone or reproducer.

S1 is a 2-pole 2-way spring loaded switch, and normally will complete circuits for listening. In these circumstances, with the extension lead from the remote unit plugged into the jack socket J1, the circuit is completed to T1 primary via contacts B. At the same time, contacts A complete the circuit from the output transformer secondary (T2) to the master unit speaker. Incoming signals are thus amplified, and heard in the master unit speaker.

When S1 is pressed, the master unit speaker is switched to T1, and audio output from T2 goes via contacts B and J1 to the remote speaker. So sounds picked up by the master unit speaker are amplified and made available for the remote loudspeaker.

Extension unit

The circuit for this is shown in Fig. 10.3. Switch S3 is normally open, and requires no manipulation during conversation. With the remote loudspeaker unit operating either as microphone or as speaker, audio signals are passed by C7. This capacitor blocks direct current, but can be shorted by the push switch S3.

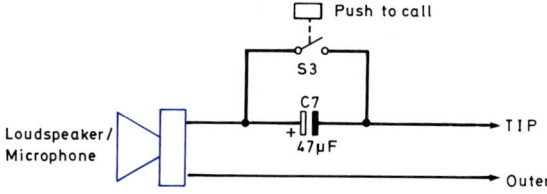

Figure 10.3

Extension unit circuit. C7 blocks d.c. but not audio signals

When S3 is pressed, the circuit is completed from battery negative (Fig. 10.2) through the primary of T1, contacts B (which are closed), and the remote speaker. This brings the amplifier into use in an unstable condition so that it oscillates, and the sound warns the master unit controller that communication is wanted. The person at the master unit then closes the main battery switch S2, and presses S1 to talk, releasing it to listen.

Should the person at the master unit wish to communicate, he closes S2, presses S1, and speaks into his microphone.

With all circuits of this kind simultaneous 2-way conversation is not possible, but this will be found of little practical disadvantage provided each person says 'go', 'over' or similarly lets it be known when he has finished. The person at the master unit can then press S1 to speak and release it to listen, allowing a smooth flow of conversation each way, as required.

Construction

The positions of components on the board can be seen from Fig. 10.4. The board is about 10 × 3cm (0.15in matrix). Several points should be noted: The three transistors each have the centre lead crossed over, so that emitter, base and collector wires come as shown. After soldering, check that base and collector wires are clear of each other.

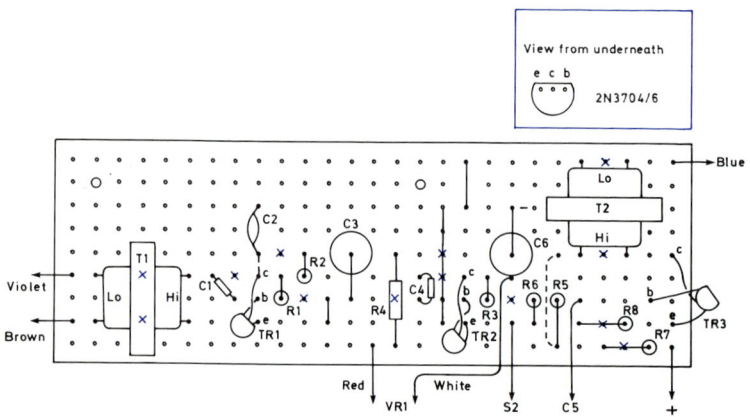

Figure 10.4
The layout of the board. The foil breaks are listed in the text

Transformers T1 and T2 are both the same, being of the type used in small radio receivers to couple a push-pull output stage to the loudspeaker. The ratio is about 6 + 6 : 1, and the centre tap is not used. With T2, the transformer is connected in the usual way, with primary to TR3, and secondary to speaker circuit. In the case of T1, the windings are reversed, the original primary being employed as secondary. So in each case the low and high impedance windings are used as indicated.

Several foil breaks are necessary, as follows.
Between low resistance tags of T2.
Between C2 and R2, between R4 and C6, and between high resistance tags of T2.
Between low and high tags of T1, between C1 and C2, between C4 and TR2 collector C.
Between R1 and lead connecting C3 negative to negative line, between C3 negative and C4, and between R3 and R6.
Between low and high tags of T1, and under R8.
Under R7.

The use of colour coded leads, as in Fig. 10.4 will allow easier wiring of external circuits, and especially those to S1.

Take care that no short circuits are allowed to arise when fixing the board in the metal case, except for the negative or chassis line. The board is easily fitted with two short 8 B.A. or similar screws, with spacers or extra nuts and washers below it.

It is easier to complete connections to VR1, S2 and S1 in particular, before fitting the board in the case.

C5 is connected directly to VR1 centre tag.

Further wiring

Fig. 10.5(a) shows actual connections to the switch S1. If an alternative type of 2-pole 2-way switch is used wire this appropriately. Leads forming any part of the input circuit to TR1 should be reasonably clear of those from the output circuit from TR3. Screened connections were not found to be necessary, wires instead being as short and direct as can be conveniently arranged, while running close to the metal case.

Fig. 10.5(b) shows connections to VR1. A separate *on-off* switch is preferable to a combined volume/*on/off* control, so that the gain can be left set at a suitable level, and so that the calling system described earlier can operate.

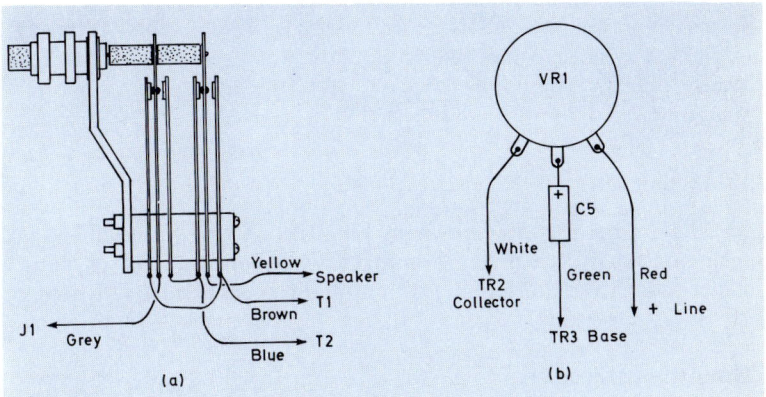

Figure 10.5

Wiring detail for S1, and connections to volume control VR1

Should an insulated case be used, it will be found necessary to earth the fixing bush of VR1, switch mounting bush, and loudspeaker frame, for best stability.

The speaker, switches and VR1 are mounted on the front of the master unit case, and the amplifier is bolted to the bottom. There is ample space in a case of the size shown for any usual type of 9V battery (Fig. 10.6).

Figure 10.6

Mounting the components in the master unit case

The socket J1 will be found useful, as it allows the extension circuit lead to be unplugged when fitting up the units. A screened lead, with the outer forming the chassis return, prevents unwanted pick-up by this circuit, but an unscreened twin conductor was also found satisfactory. Correct polarity should be checked with the latter, because of the capacitor C7.

Remote unit

Connections for this are shown in Fig. 10.7. S3 is the usual type of spring loaded switch making contact when pressed. A very much smaller case than required for the master unit is easily adequate here.

Using the intercom

The method of operation has been described. It is worthwhile placing instructions at each switch, so that anyone can use the equipment readily. S1 can be marked *Push to Call or Talk*. S2 is marked *on-off* and S3 *Push to Call. Release to Talk and Listen.*

It will generally be found that VR1 should not be set for maximum amplification.

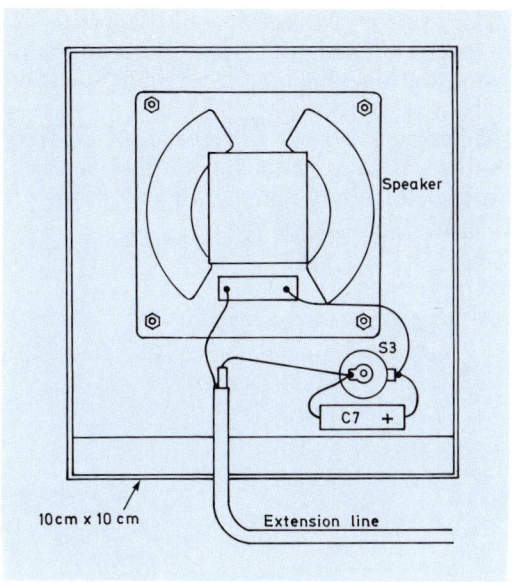

Figure 10.7
Connections and components in the small case

Troubleshooting

In the event of no results at all being obtained, the amplifier can be checked by temporarily taking one speaker to T1, and the other to T2, thus eliminating the switching associated with S1. If amplification of sounds is then obtained, look for some fault in the wiring or operation of S1.

When first testing the equipment, do not forget that sounds from one speaker reaching the other will cause continuous howling feedback. If there is other than a negligible battery current with S2 in the off position, C7 should be changed.

Table 10.1. Components list for home intercom

Resistors	
R1	1.8MΩ (all resistors 5%, ¼W)
R2	10kΩ
R3	1.8MΩ
R4	3.9kΩ
R5	22kΩ
R6	12kΩ
R7	150Ω
R8	18Ω
VR1	5kΩ log potentiometer
Capacitors	
C1	4.7nF
C2	10nF
C3	100μF, 10V
C4	47nF
C5	1μF, 10V
C6	100μF, 10V
C7	47μF, 10V
Semiconductors	
TR1	2N3706
TR2	2N3704
TR3	2N3704
Switches	
S1	2-pole 2-way spring loaded
S2	*on-off* switch
S3	push-for-on switch
Miscellaneous	
T1 and T2	coupling transformers, ratio about 12 : 1
	two small loudspeakers, approximately 6cm diameter, 2Ω or 3Ω
	sloping front case about 20 × 15 × 15cm
	sloping front case about 10 × 10 × 10cm
	board approximately 10cm × 4cm (0.15in matrix), jack socket, knob, plug and extension lead, battery clips, etc

11

Metal Locator

A metal detector responds to metal even when buried underground, and it can be used for searching for all sorts of things. In some cases valuable

Figure 11.1
The finished metal detector

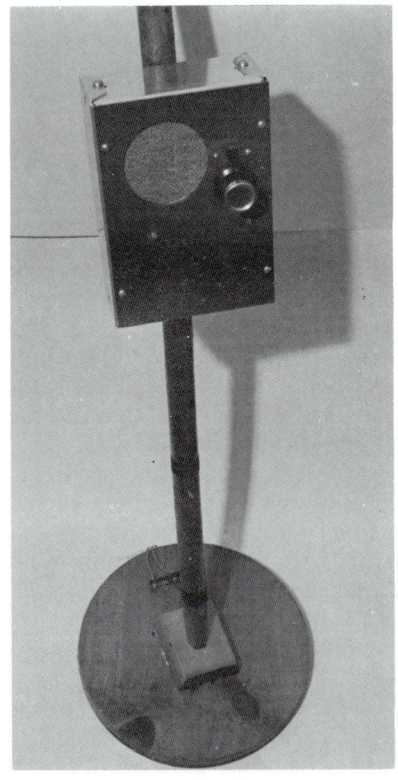

objects have been found with the aid of such an instrument, though the richest finds have probably been also due to a knowledge of likely sites, as well as some luck. It is not to be supposed that most buried or lost objects are valuable, but despite this many people find interest in trying to find valuable items. Such an instrument may also have more ordinary uses, such as the location of concealed or buried drain access covers and similar hidden works.

The circuit

Various circuits may be used, and the beat frequency oscillator (BFO) type of detector (Fig. 11.2) is popular for home construction. It has relatively simple circuitry, is easy to get working, and can give a useful performance. Many of the less expensive commercially manufactured locators are of BFO type; they can give a good performance, and avoid some of the snags of induction balance and pulse locators. The locator described here is of BFO type.

The instrument responds only to metal objects. The depth at which these can be located depends on the type of soil, size and kind of object, and similar factors. This locator gives the same performance in this respect at ready-manufactured BFO locators.

Search oscillator

L1 in Fig. 11.2 is the search coil. It is 20cm in diameter, of flat construction, and is moved about near the ground. The frequency of operation depends primarily upon the coil, and the parallel capacitances C1, C2, C4 and C5. C2 is a trimmer for adjustment of frequency, as described later.

TR1 is the search coil oscillator, and the emitter tap provided by C4 and C5 in series results in the correct phase of feedback for oscillation, without a tapping or two windings at L1. Base current is obtained through R1. R3 and C6 decouple this oscillator and avoid unwanted frequency pulling effects with the reference oscillator TR2.

L2 is a small coil with adjustable core, the exact frequency being largely determined by the core position and parallel capacitor C9. Feedback to the base of TR2 is via C8. The tapping 2 on L2 provides the correct phase. After initial adjustment, the frequency here can be controlled by the variable heterodyne capacitor VC1, which is manually operated.

To conform with licence requirements, both oscillators work at about 95kHz. The small capacitors C7 and C10 couple the two oscillators to the heterodyne detector circuit D1 and D2.

If TR1 and TR2 are operating on exactly the same frequency, no heterodyne is produced, so there is no audible signal. However, when the oscillators are on slightly different frequencies, the difference is

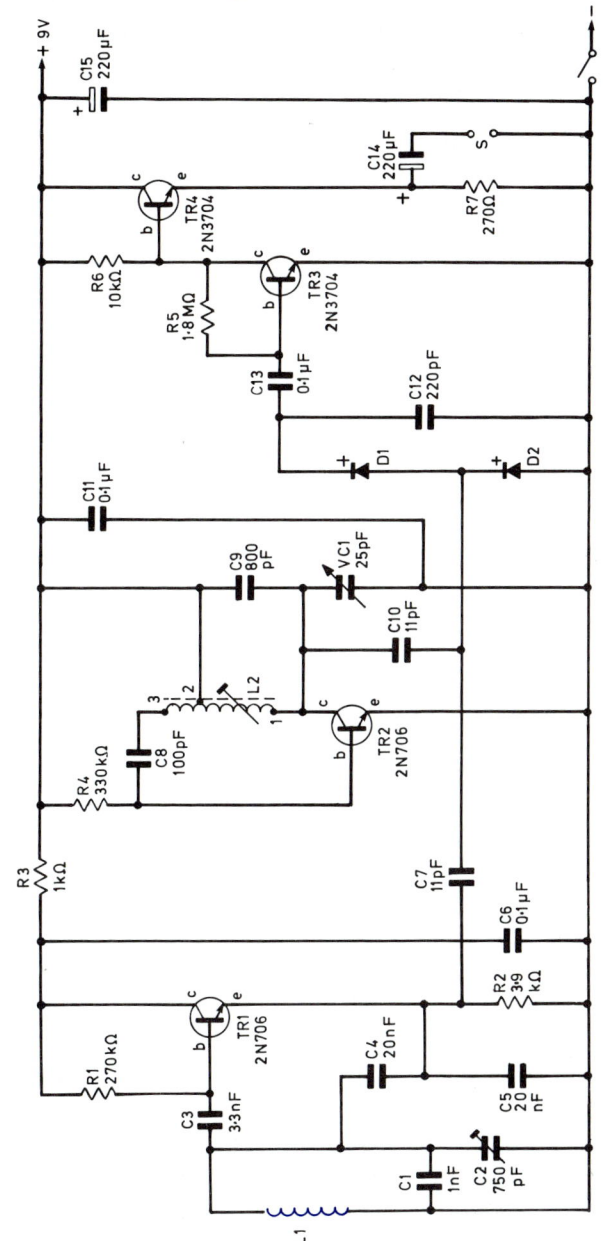

Figure 11.2

Circuit of the metal detector. L1 is the search coil

heard as an audible heterodyne. As example, if TR1 operates at 95kHz (95,000 hertz) and TR2 at 96kHz (96,000 hertz) an audible tone of 1kHz would be obtained from the heterodyne detector. So the audible tone is absent with both oscillators working at the same frequency, but appears and rises in pitch in accordance with any increasing difference in frequency between the oscillators.

The frequency of the reference oscillator TR2 is left unchanged, after initial setting by VC1. When metal comes within the field of the search coil L1, the frequency of the oscillator TR1 is changed, so producing an audio tone, or changing the tone originally present.

Audio section

TR3 and TR4 form a simple direct coupled audio amplifier, with good gain and enough output to drive a small loudspeaker. R5 supplies base bias to TR3. Operating conditions for TR4 are controlled by TR3, R6 and R7, and do not depend on the speaker load, as this item is coupled by C14.

A speaker of about 75Ω is most suitable, though 40Ω is satisfactory. In some circumstances (such as high wind, or with audible signals being a nuisance to others) headphones may be preferred. Medium impedance and similar headsets can be plugged in if wished by having a jack socket for the speaker circuit. This can be wired to switch off the speaker, when the plug is inserted.

The audio gain present, and volume, are matters primarily of convenience in use, as they have no bearing on the operation of the heterodyne circuit, or the sensitivity of detection.

Search coil

This coil has 20 turns of 32swg enamelled wire wound on a 20cm diameter former in a 5mm wide slot. If tools are available for the job,

Figure 11.3
The search coil is wound on a wooden former

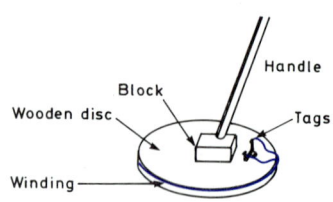

a 5mm slot can be cut in a disc of wood about 10 to 13mm thick and 21cm diameter. An alternative is to make the disc from 4mm hardboard.

This will require three wooden discs — one 20.5cm in diameter, and two 21cm in diameter. The smaller disc is cemented between the other two, and the adhesive is allowed to harden with the whole clamped together, or under weights, so that there is no gap into which the wire can slip when winding. Any type of former should be varnished before use, to keep out damp. Shellac or other outdoor varnish is suitable.

The ends of the winding are soldered to two small tags or a 2-way tagstrip. Also cement on a central block of wood, with a hole to take a handle, as in Fig. 11.3. A piece of broomstick, about 1m long, can be tapered so that it is a push fit in the wooden block.

Other search coils may be made, but turns must be so arranged that it is possible to tune the search coil circuit in the way explained later.

Reference oscillator coil

This is coil L2 in Fig. 11.2. It has 680 turns of wire in all, on a 7mm diameter former with an adjustable core. The wire is 34swg enamelled. All turns are in the same direction and there are 600 turns from 1 to 2 (see Fig. 11.2), and 80 turns from 2 to 3.

Figure 11.4
The reference oscillator coil

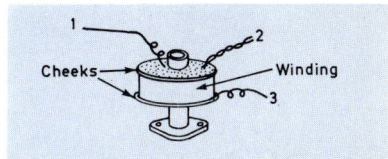

Prepare the former by fitting on two discs, each about 22mm in diameter, located as in Fig. 11.4. A winding space of 9mm is left between them, and they can be made from any strong and smooth insulated material, and are cemented in position.

Drill a small hole adjacent to the 7mm former, pass the wire through so as to leave a few centimetres projecting (1), and wind the 600 turn section. Form a loop and bring it out through a hole (2), and wind 80 more turns in the same direction, ending at the outside of the coil (3). Put a layer of adhesive tape over the coil.

A slight error in the number of turns is not important. Winding can be speeded by using some mechanical means, such as a geared hand drill. Clamp the drill in a vice, and fit the coil former over a bolt or threaded rod held in the drill chuck. The drill handle can then be rotated with one hand, while the other hand guides the wire. Count the gear ratio before starting, to allow for this. Suppose it is 4 : 1. The 600 turn section will require 150 revolutions of the handle, and the 80 turn section 20 revolutions.

Circuit board assembly

The board should be approximately 8cm × 6cm (Fig. 11.5). When fitting the trimmer C2, take care that this does not short circuit any foils. Place an insulated washer under the securing nut, or cut breaks

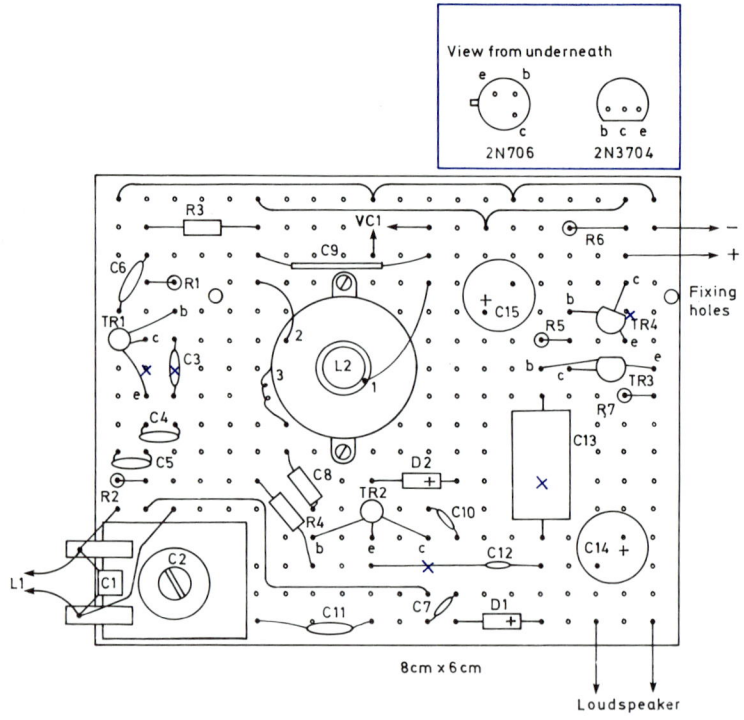

Figure 11.5

The board and components on it. C2 is a trimmer to set the search coil frequency

to isolate this. If L2 is fitted with 8 B.A. bolts and nuts, these will not bridge adjacent foils. Otherwise take similar precautions.

Foil breaks are required between emitter and collector of TR1, under C3, between TR2 collector and C7, under C13, and between emitter and collector of TR4.

Provide red and black leads for the battery (via switch), and connections for the speaker (C14 and negative) and VC1. With the latter, the fixed plates tag will go to 1 at L2, and rotor connection to negative.

Tuning

It is best to test and provisionally tune the oscillators before complete assembly. The leads from L1 are soldered to C2.

No audio output can be produced unless both oscillators are operating on nearly the same frequency. Initially, tune these to 100kHz, as the second harmonic of this frequency can be picked up with a portable or other radio having a long wave band, and tuned to the 200kHz programme. With the programme tuned in and the receiver near the board, adjust C2 until a strong whistle is heard accompanying the programme. Then slightly screw down C2 until this has just ceased to be heard.

Proceed in the same way when adjusting the core of L2. This will place both oscillators a little on the low frequency side of 100kHz and a heterodyne will probably be heard in the locator speaker. When this is so, adjust C2, and the core of L2, so that the 'zero beat' position (no whistle) is obtained with VC1 half closed. In these circumstances, a heterodyne tone will be heard, and will rise in frequency, when VC1 is either opened or closed from its middle position.

If no heterodyne can be obtained, the oscillators may be too far off frequency to allow adjustment, or one or both may not be oscillating. As C2 and L2 can be adjusted over a wide range, it is unlikely that the oscillators cannot be shifted to a suitable common frequency.

If it is necessary to check if TR1 is oscillating, take a connection from R3, temporarily lifting the positive end of this out of the board. Place a multi-range testmeter or similar instrument in series with the battery supply to TR1 only. Current should change when C2 is temporarily shorted. If it does not, TR1 is not oscillating. Possible causes include overlooked joints, a break in L1, or defective transistor.

A similar test is readily made with TR2, if necessary. To do this, temporarily join 2 of L2 and C9 above the board, and insert the meter between this point and positive. If current changes when C9 is shorted, TR2 is oscillating.

Should it be necessary, the audio section can be checked by connecting medium or high impedance phones across C12. If the heterodyne can be heard here, but not from the speaker, then a check needs to be made of connections and components in the audio amplifier C13, R5, R6, TR3 and TR4, R7 and C14.

Construction

A metal or plastics box large enough to hold the circuit board, battery and speaker will be most convenient (Fig. 11.6). VC1 and the *on-off* switch can be fitted to the box, which is mounted about half way up the detector handle. Run leads from C2 through a small hole, and down to the search coil tags. Tape these leads to the handle. A cycle grip or similar item may be fitted to the top of the handle.

Figure 11.6

Assembling the components in the box

An aperture must be cut, or a number of smaller holes must be drilled, to match the speaker cone. If the brittle type of plastic box is used, tools must be sharp, and applied with light pressure, otherwise the material may crack.

The whole assembly should be reasonably rigid, and preferably so made that rain cannot easily enter. Unnecessary weight is best avoided.

Using the detector

Maximum sensitivity is obtained if VC1 is set so that a low pitched audio tone is produced. This will then change in frequency when the search coil is brought near a metal object. If the tone drops in frequency and ceases, then rises, as metal is approached, set VC1 on the opposite side of the 'zero beat' position described.

A few tests will show how the device operates, and how it should be adjusted and used for best results.

With wet sand, earth, and similar conductive but non-metal bodies, a capacitance de-tuning effect can arise. This can be greatly reduced or avoided by using a Faraday shield for L1. The shield is most easily made from aluminium household foil, cut to match the winding L1,

Table 11.1. Components list for metal detector

Capacitors

C1	1nF 5%
C2	750pF or 1nF trimmer
C3	3.3nF
C4	20nF, 5%
C5	20nF, 5%
C6	0.1μF
C7	11pF
C8	100pF
C9	800pF, 5%
C10	11pF
C11	0.1μF
C12	220pF
C13	0.1μF
C14	220μF, 10V
C15	220μF, 10V
VC1	25pF or similar small variable capacitor

Resistors

R1	270kΩ (all resistors 5%, ¼W)
R2	3.9kΩ
R3	1kΩ
R4	330kΩ
R5	1.8MΩ
R6	10kΩ
R7	270Ω

Semiconductors

TR1	2N706
TR2	2N706
TR3	2N3704
TR4	2N3704
D1 and D2	1N67A, 0A81, etc

Miscellaneous

L1 and L2	see text
	0.15in matrix board approximately 8cm × 6cm
	6cm, 75Ω or similar speaker
	knob, battery clips
	on-off switch, broomhandle, battery clips, 15 × 10 × 5cm insulated box

and folded around it. This can be done only be removing L1 from the former. L1 must thus be wound on a disc or similar object without a groove, so that it can be carefully taken off, and bound with thread to keep the turns together. The shield can then be fitted. The coil, with shield, can be cemented to a disc of hardboard about 21.5cm in diameter. A gap of about 3mm must be left in the shield, and it is connected to L1 at the negative end of the winding. Re-trimming will be necessary, due to the shift in frequency. In some cases it may be necessary to change C1, to obtain the correct frequency, depending on how closely the foil shield surrounds the turns of L1.

12

Process Timer

This timer has a delay that is adjustable from approximately 5 seconds to 4 minutes, and the latter interval can easily be extended, if required. In operation, when its switch is set to 'time' and the selected interval has passed, an external circuit will be switched on, or off, as wished.

Figure 12.1
The process timer in its case

An internal audible warning is incorporated in the timer, and this is useful for processes which will be ended manually.

The external circuit can control a bell, buzzer or oscillator if an audible warning is wanted at a distance; or will switch the required equipment directly if necessary. Various switching and external control arrangements can be adopted, depending on the way in which the timer is to be used. It may, as example, control an enlarger, when making a number of prints with the same exposure.

The circuit

Fig. 12.2 is the complete circuit of the timer, and can be looked upon as two separate sections — the timing circuit itself, and the audio

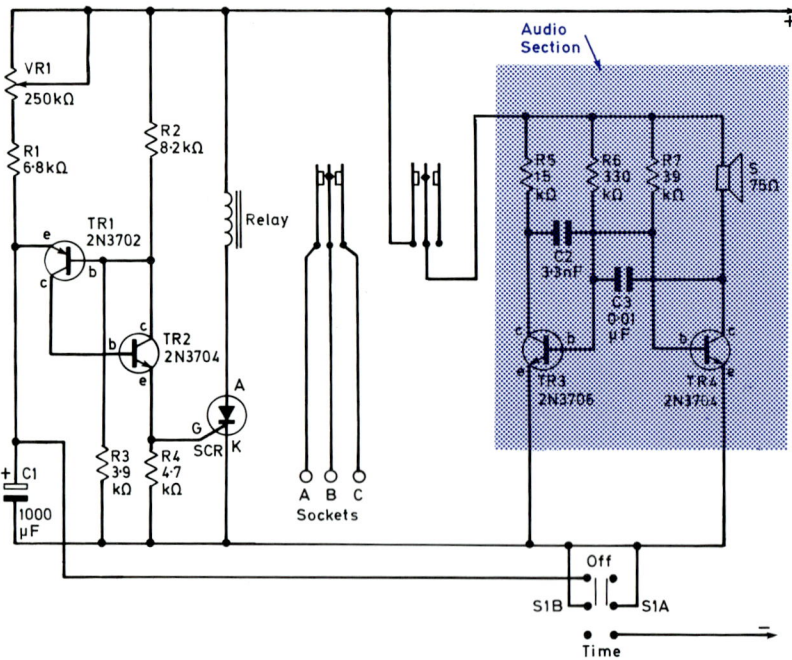

Figure 12.2

The timing circuit operating relay. TR3 and TR4 produce the audible warning signal

oscillator. These are in fact constructed on individual boards. It is convenient to operate the unit from a 9V battery, but an external 12V or similar supply can be used instead.

The basis of the timing circuit is the charging of capacitor C1, and the operation of the Schmitt trigger TR1 and TR2.

S1A and S1B are the two poles of the 'off' and 'time' switch. This switch is in the 'off' position in Fig. 12.2. S1A disconnects the negative supply, while S1B short circuits C1, to discharge it, so that timing intervals can be repeated.

When the switch is moved to 'time' the negative circuit is completed and S1B opens, so that C1 begins to charge. Initially, the base potential of TR1 depends on the potential divider formed by R2 and R3. Since C1 is discharged, TR1 base is positive relative to its emitter. TR1 is a pnp transistor, so negligible emitter and collector currents flow.

Eventually C1 charges to a potential (about 4V to 5V) where the emitter of TR1 is going to be positive relative to the base. This causes TR1 to conduct due to the negative base bias present. The base of TR2 is thus moved positive, as it is common to the collector of TR1 and TR2 conducts. When TR2 starts to conduct, collector current through R2 increases the voltage drop in this resistor, thus moving TR1 base further negative. A result of this is that when the critical emitter voltage of TR1 is reached, the effect is instantly cumulative with TR2 passing emitter current through R4. This current produces a voltage drop in R4, moving the gate of the SCR positive, so that anode current passes. This energises the relay.

In the relay, contacts connected to sockets A and B are normally open, and B and C are normally closed. It is thus possible to complete the external circuit by connecting to A and B, or to interrupt it by connecting to B and C.

The second set of contacts close to apply voltage to the audible warning board. This consists of the multivibrator TR3 and TR4, the latter operating a small internal loudspeaker. The working of this type of audio oscillator has been described earlier in Project 3.

Construction

The timer circuit board is about 6.7cm × 7.5cm (0.15in matrix) and components are fitted as in Fig. 12.3. No breaks are required in the foils.

A relay which plugs into an octal type holder is convenient but other relays will be satisfactory. The relay coil is for 12V working, and has a resistance of about 120Ω. A relay of this kind will normally operate satisfactorily with a somewhat lower voltage, and is suitable for 9V. With other relays, a coil resistance of about 100Ω to 250Ω is most suitable. A very low resistance coil is unsuitable because the current taken will be too large.

A hole has to be cut for the octal holder, or to clear the projecting tags of the relay, if of the fixed type. This can be made by drilling a ring of small holes close together, carefully breaking out the piece, and finishing off with a round or half-round file. Two 6 B.A. bolts and nuts fix the holder.

Solder on two leads to go to VR1, and also connections for the switch, as shown. Make the usual check to see that connections are correct, and that no fragments of solder bridge adjacent foils. Two small brackets are used to mount the board. Assure that the fixing screws or nuts for these do not cause any short circuit.

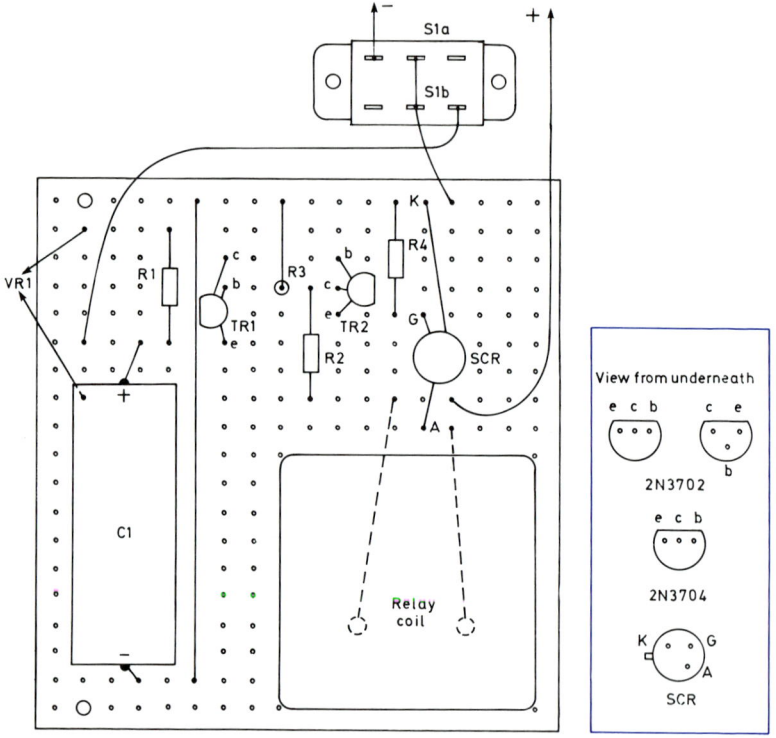

Figure 12.3

There are no foil breaks on the circuit board

The oscillator board is shown in Fig. 12.4. First cut a hole about 4½cm in diameter for the speaker, and cement it in position. Foil breaks are required under C2 and C3, and also to isolate the one fixing bracket. Note how connections from C2 and C3 cross over.

Interconnection

Both boards are mounted by means of small brackets to the panel, which is 15cm × 10cm (as in Fig. 12.5). The fixing brackets for the audio board (Fig. 12.4) are positioned so that they can be secured by the two bolts which fix the switch to the panel.

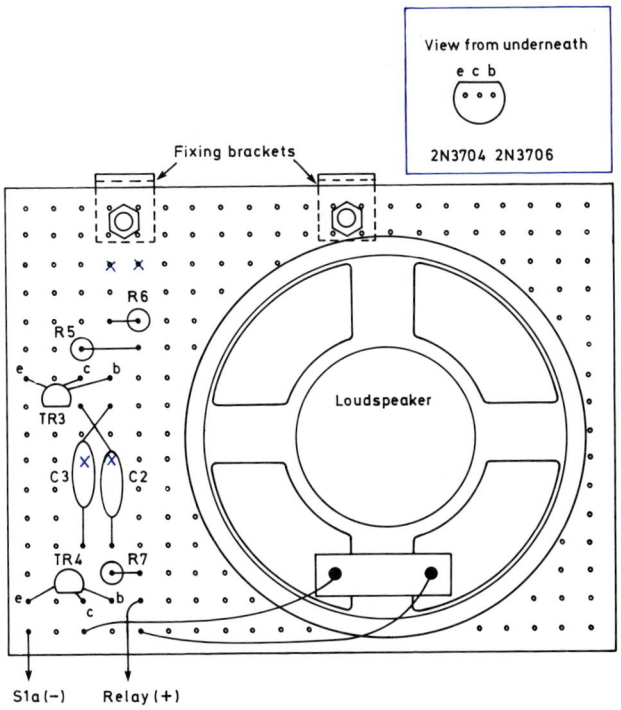

Figure 12.4

The audible warning board with the loudspeaker

The negative lines of both boards are connected together. The audio board positive line is connected to the relay contacts, which will complete the positive circuit when the relay is energised.

Calibration

The exact times obtained for various settings of VR1 will depend on component tolerances. Only a slight change is caused by modifying the supply voltage. However, timing is best done with the intended supply — an internal 9V battery, or 12V externally provided.

With VR1 out of circuit, a delay of about 5 seconds should be expected, rising to some 4 minutes or so with VR1 adjusted so that all its element is in circuit. Fit a scale to VR1 and calibrate this by means of a watch with a seconds hand.

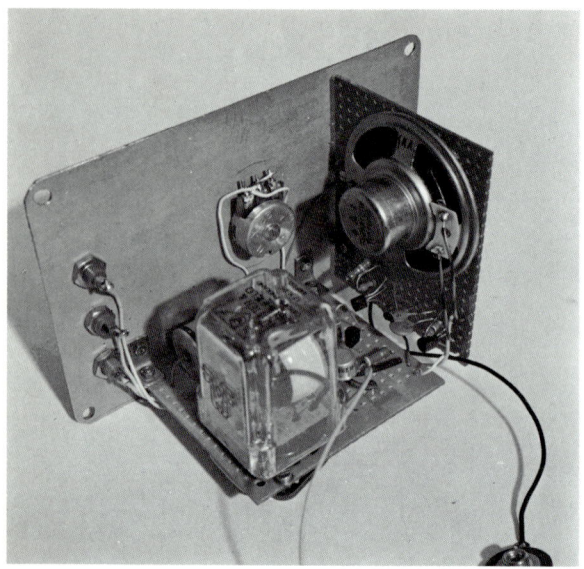

Figure 12.5

The boards mounted on the panel

The longest interval obtainable can be increased by having a larger capacitance at C1, or by increasing the value of VR1. This may be preferred for some purposes. As a guide, C1 of 4700µF provided a delay of up to 12 minutes.

Should the circuit fail to operate, especially at the more lengthy interval settings of VR1, and no fault can be found, then it may be necessary to change C1. If this component is leaky, it may not charge to the point required to trigger TR1 and TR2. With a leaky capacitor, the circuit may still operate with VR1 set for short intervals. C1 should be of good quality — not ancient surplus.

Case

A case 15 × 10 × 10cm will house the timer, and can be metal or plastics. The three insulated sockets A, B and C are fitted to the panel. A 9V battery can be fitted between the audio board and side of the case.

External circuits

The maximum voltage and current which can be switched by the relay contacts will depend on the contact ratings. A good quality octal relay can be expected to be available with ratings of up to 3A 250V a.c., 2A at 24V d.c., or 50mA at 250V d.c.

Small low voltage relays, as used for various projects, and not intended for mains voltages, must not be used with high voltage.

The sockets on the case are intended for circuits where up to about 24V maximum will be present — low voltage lamps, model motors, trains, and similar items.

In the interests of safety, properly insulated and permanently connected circuits must be used, for mains voltages.

Where some unusually heavy, high voltage load has to be controlled, a secondary relay should be used. This will have contacts and insulation to suit the load, and its coil will be energised by a 12V or 24V circuit from the relay in the timer.

Checking

If the trigger section does not work, use a *high resistance* voltmeter to find if there is a slow rise in potential across C1. (This may prevent the

Table 12.1. Components list for process timer

Resistors
R1	6.8kΩ (all resistors 5%, ¼W)
R2	8.2kΩ
R3	3.9kΩ
R4	4.7kΩ
R5	15kΩ
R6	330kΩ
R7	39kΩ
VR1	250kΩ linear potentiometer

Capacitor
C1	1000μF, 10V

Semiconductors
TR1	2N3702
TR2	2N3704
TR3	2N3706
TR4	2N3704
SCR	silicon controlled rectifier, 50V, 1A

Miscellaneous
6cm 75Ω or similar speaker
S1A/S1B 2-pole 2-way slide switch
relay, octal plug-in, 12V, with 2-pole 2-way change over contacts. Holder for same
case about 15 × 10 × 10cm
board (7.5 × 6.7cm) 0.15in matrix
board (8 × 6.5cm) 0.15in matrix
three sockets, knob, etc

timer working at *lengthy* intervals.) If not, check connections to the switch, VR1, and also R1. With this in order, an abrupt rise in voltage should be found across R4. If not, R2, R3, R4 and TR1 and TR2 should be investigated.

Momentarily connecting TR2 collector to the SCR gate should trigger the latter so that anode current passes to operate the relay. (The SCR remains in this condition until S1A is opened.) If the relay can be heard or seen to operate check if necessary that the correct tags or pins are used for A, B and C.

The audio board can be checked, if necessary, by taking power to it directly. Here, a wide range of component values and transistors can be used. These will, however, affect the note produced.

If other transistors are tried in the TR1 and TR2 positions, note that TR1 is a *pnp* type, but TR2 is *npn*.

13

Reaction Indicator

You can use this device alone, or in conjunction with one or two other players. Its main purpose is to give an indication of the relative speed with which a person can respond to the appearance of a signal,

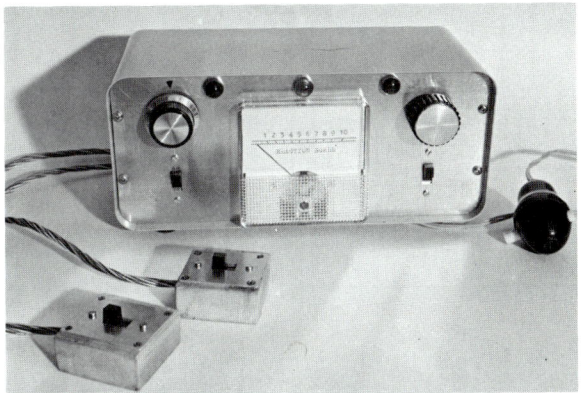

Figure 13.1
The reaction indicator

and with two persons competing it also shows who was first. The signal may be initiated automatically, or can be provided manually by a third person.

The unit has a priority indicator system which may be used for other games — such as 'snap' and quiz contests. It can thus afford considerable entertainment in several different ways.

The circuit

Fig. 13.2 is the complete circuit, and the silicon controlled rectifier and associated items form the delay or time part of the equipment. The

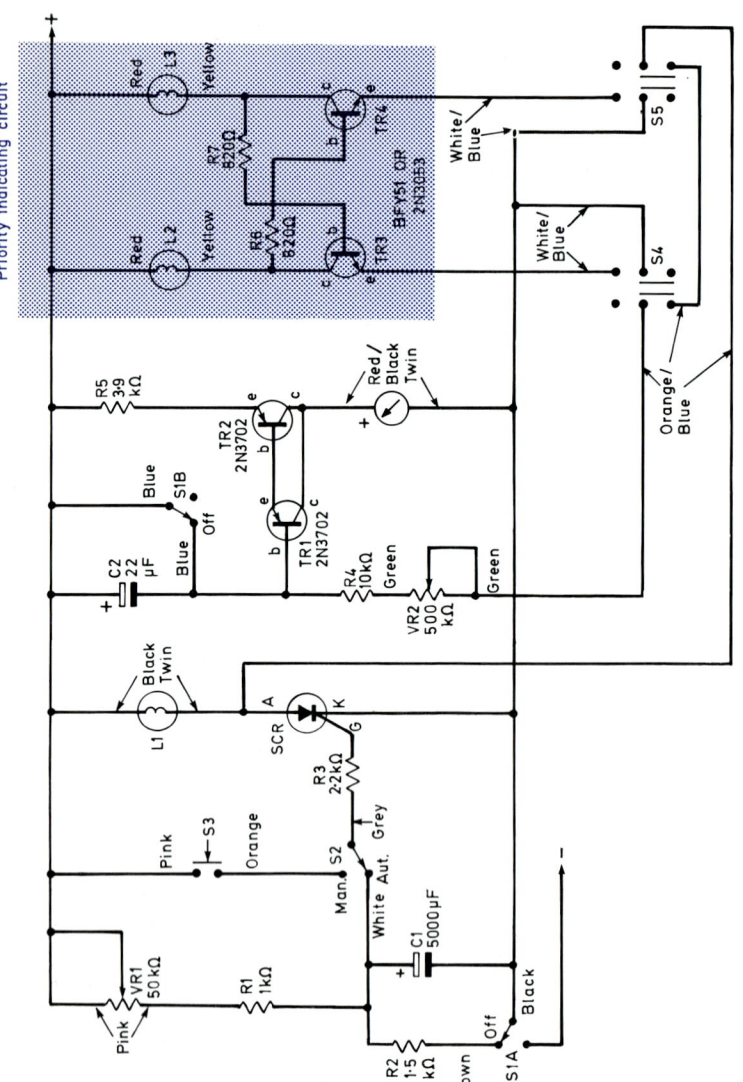

Figure 13.2

The indicator has time/manual, lapse indicator, and priority circuits

purpose of this is to illuminate the bulb L1 with a delay of about 1 second to 30 seconds. When L1 lights, the competitor with the faster reaction wins.

Switch section S1A is the main *on-off* switch. When in the 'off' position C1 is discharged through R2, this discharge being complete if S1A is left 'off' other than briefly. With S1A 'on', C1 commences to charge through R1 and VR1. The exact length of charging time is not known by the competitors, but can be changed by means of VR1.

When S2 is at 'automatic' (AUT.), the gate G of the SCR moves positive until a level is reached where the SCR is triggered into conduction. Avalanche current lights L1, and the voltage across the SCR falls to a low level.

If S2 is at 'manual' (MAN.), the triggering of the SCR is under the control of the switch S3, which is operated by a third person. This provides further scope for competition.

After each attempt by the competitors, S1A will be returned to the off position, so that the sequence described can be repeated.

Lapse indicator

This consists of TR1 and TR2, with associated circuitry. The switches S4 and S5 will normally be in the position shown. So when the SCR is triggered and its anode moved abruptly negative, a negative polarity supply becomes available, through VR2 and R4, to charge C2.

As C2 charges, a negative voltage arises on the base b of TR1 of the high-gain pair TR1/TR2. As these are *pnp* transistors, TR2 commences to pass a rising collector current, shown by the meter. The speed with which C2 charges, and hence the speed with which the meter pointer moves across the scale, can be set by VR2. The most rapid movement allows a clear indication of brief intervals, as when the competitor's reaction arises in a small part of a second. The slower rate allows some seconds, so that it can be used for competitive attempts which will require such intervals.

S4 and S5 are the individual competitor's switches. Either will break the circuit from the SCR anode to C2, so that the latter stops charging. As there is now no discharge circuit except through TR1, and base current here is extremely low with the high-gain pair, the meter indication remains with negligible falling off, and can be noted.

Switch S1B is the second pole of the main 'off' switch, so that when S1A/S1B is put into the *off* position, C2 is completely discharged, so that the timing interval shown by the meter can be repeated.

In play, when L1 lights, each competitor reacts as rapidly as possible, operating switch S4 or S5. The meter pointer commences to rise when L1 is illuminated, so the most rapid reaction will be that which halts the meter pointer at its lowest position.

Priority circuit

It is clear that both competitors will probably respond within a small fraction of a second, so that it is necessary to know who was first. TR3 and TR4, with the additional poles of the switches S4 and S5, form the priority indicating circuit.

With both emitter circuits open (neither switch operated) base current for either transistor can be obtained from R6 or R7, as the collectors are fully positive.

Assume S4 is operated first. TR3 receives positive bias from R7, through L3, so passes collector current to light bulb L2. If S5 is now closed, voltage drop in L2 has resulted in the collector of TR3 being so negative that R6 cannot supply base current for TR4, so that L3 does not light. In the same way, if S5 is operated first, L2 cannot be lit by closing the emitter circuit to TR3.

Lamps L2 and L3 therefore show which switch was operated first, even though the competitors' reactions may have seemed to be simultaneous.

The whole unit could be self-contained with S3, S4 and S5 mounted on the case. But it will be found better in practice to have these switches on flexible cords, with the main switch, VR1, VR2, and the indicator lamps fitted to the case.

Construction

The circuit board is about 10.5cm × 5.5cm (0.15in matrix) and is shown in Fig. 13.3. As components and leads are soldered in position, foil breaks are made as follows:

> between R1 and positive line
> between R3 and TR2 emitter, TR2 emitter and R7
> between TR2 base and R6
> between TR2 collector and TR4 emitter
> under R4 and between 'greens'
> between SCR anode and TR3 emitter
> between SCR gate and 'orange/blues'

Colour coded leads will facilitate the connecting up of the switches and other peripheral components. It is also possible to wire these, without colour coded connections, as construction proceeds. The following will be noted: pinks near R1 run to VR1; blues to S1B allow shorting of C2. A black twin runs to L1, with yellow and red to L2, and to L3. The red-black twin is for the meter. Greens run to VR2. Grey to S2 allows switching to C1 (white) or to manual, the orange lead running from S2 to S3.

Switches S4 and S5 each have four colour-coded leads. The white/blue complete the emitter circuits; the orange/blues are joined at the circuit board, and complete the circuit from the SCR anode to VR2, through S4 and S5 in series.

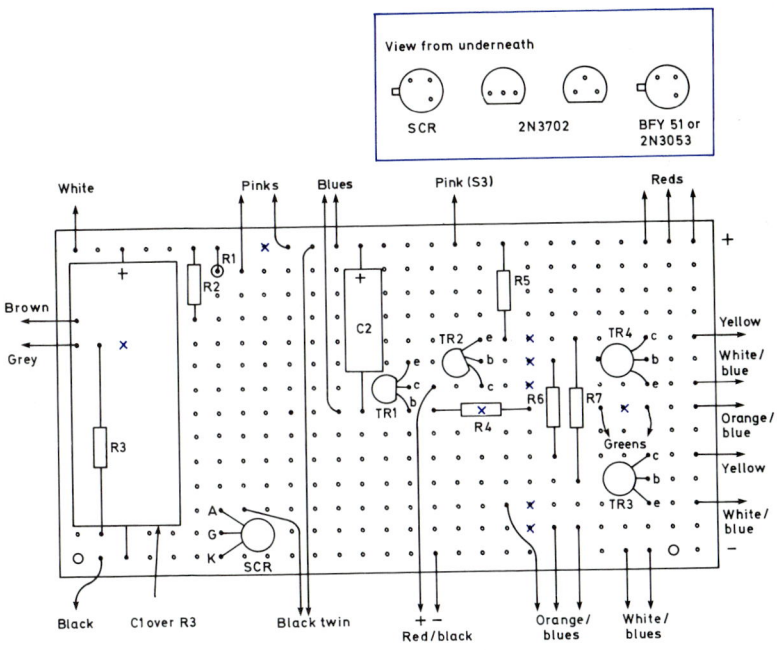

Figure 13.3

Coloured leads simplify connection of external components to board

R3 lies under C1. Alternative lead-outs are seen for the 2N3702 transistor types and these are shown. Position the emitter, base and collector leads as in Fig. 13.3.

Sectional tests

As the whole circuit divides so readily into three sections, it can be worthwhile testing each as it is completed.

With the SCR and associated components in place, and the battery connected, check that L1 lights if S2 is at MAN. and S3 is pressed. Also that L1 lights after a delay, adjustable by means of VR1, with

S2 at AUT. Should there be any difficulty here, it will be necessary to investigate only the few components so far fitted.

With TR1, TR2 and associated items added, note that the meter commences to rise when L1 lights, and that the speed of movement can be adjusted by means of VR2. Moving either S4 or S5 from the position in Fig. 13.2 should cause the meter pointer to remain where it has reached on the scale. Any fault in this section would lie around C2, R4, TR1, TR2, and associated items.

TR3 and TR4, with the lamps and R6 and R7, can then be added. It should be possible to light only L2 or L3, depending on which switch, S4 or S5, is operated first.

Competitors' switches

These are bolted to small metal plates, which in turn are screwed to wooden blocks, having cavities as in Fig. 13.4. The non-miniature type of slide switch is most suitable. These need blocks about 4½cm square, cut from 2cm thick wood.

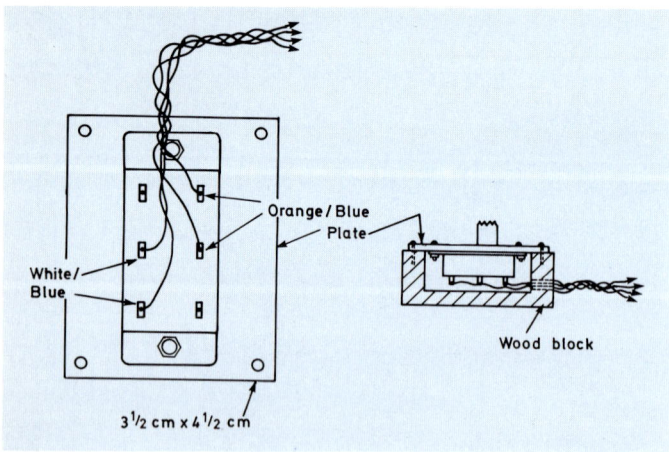

Figure 13.4

Construction details of competitors' switches S4 and S5

The four colour-coded leads are soldered to the switch, and pass through a hole in the back of the block. Four small wood screws hold the plate, with switch, in position.

Panel layout

Fig. 13.5 shows the positions of items on the 23cm × 10cm panel. To prepare this, punch holes for L1, L2, L3, VR1 and VR2, and cut a hole to clear the meter, and drill to suit its mounting studs. Apertures for the slide switches S1A/B and S2 are made by drilling several small holes, and completing with a small file. Each is held by two 5mm 6 B.A. or 8 B.A. bolts, with nuts.

Two brackets are made from scrap metal, about 45mm × 12mm, cranked to allow the circuit board to be mounted behind the meter. The meter studs hold these brackets. Use extra nuts or spacers under the board so that no short circuit is caused by the fixing.

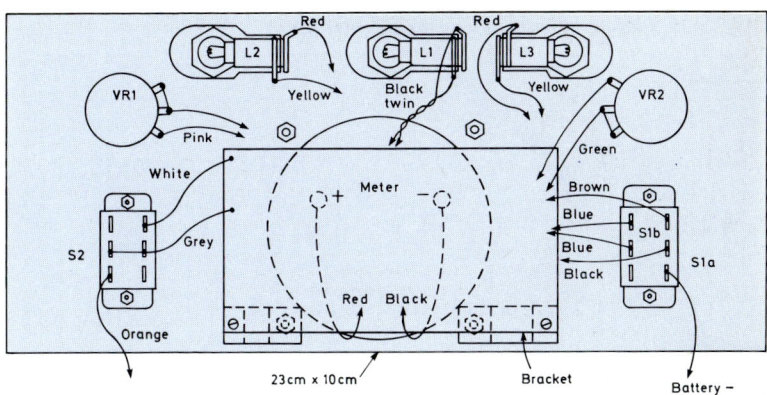

Figure 13.5
Components behind the panel

The various coloured leads can then be cut down to length, and run to the connecting points required. Other colour-coded leads, which run to S4, S5 etc., are shown in Fig. 13.5, and are connected as in Fig. 13.3.

A scale marked 1 to 10 can be made and fitted to the meter. With VR2 allowing adjustment for various games in the way described, it is not possible to fit a scale to the meter which indicates the actual lapse in time, so readings are comparative. VR2 has a calibrated knob so that settings can be repeated. A plain knob should be used with VR1.

The unit is intended to run from a 6V supply, and by fitting 0.1A bulbs, current drain is sufficiently low for battery working.

Circuit checks have been described, and faults are unlikely. If C1 and C2 are good, non-leaky components, the circuits can be expected to work correctly with no difficulty.

Table 13.1. Components list for reaction indicator

Resistors	
R1	1kΩ (all resistors 5%, ¼W)
R2	1.5kΩ
R3	2.2kΩ
R4	10kΩ
R5	3.9kΩ
R6	820Ω
R7	820Ω
VR1	50kΩ linear potentiometer
VR2	500kΩ linear potentiometer
Capacitors	
C1	4 700μF or 5 000μF, 12V
C2	22μF, 25V
Semiconductors	
SCR	50V, 1A silicon controlled rectifier
TR1	2N3702
TR2	2N3702
TR3	BFY51 or 2N3053
TR3	BFY51 or 2N3053
Miscellaneous	
L1	6V, 0.1A bulb and white holder
L2	6V, 0.1A bulb and red holder
L3	6V, 0.1A bulb and green holder
S1A/B	2-pole 2-way slide switch
S2	2-way slide switch
S4, S5	2 off 2-pole 2-way slide switches board about 10.5cm × 5.5cm (0.15in matrix), knobs, thin flex, etc
M	1mA meter (7.5 × 7.5cm instrument fitted) case approximately 23 × 10 × 10cm

Appendix

Distinguishing terminals in semiconductors

Although all transistors have terminals that are easily recognised on circuit diagrams, distinguishing them on the actual device isn't always

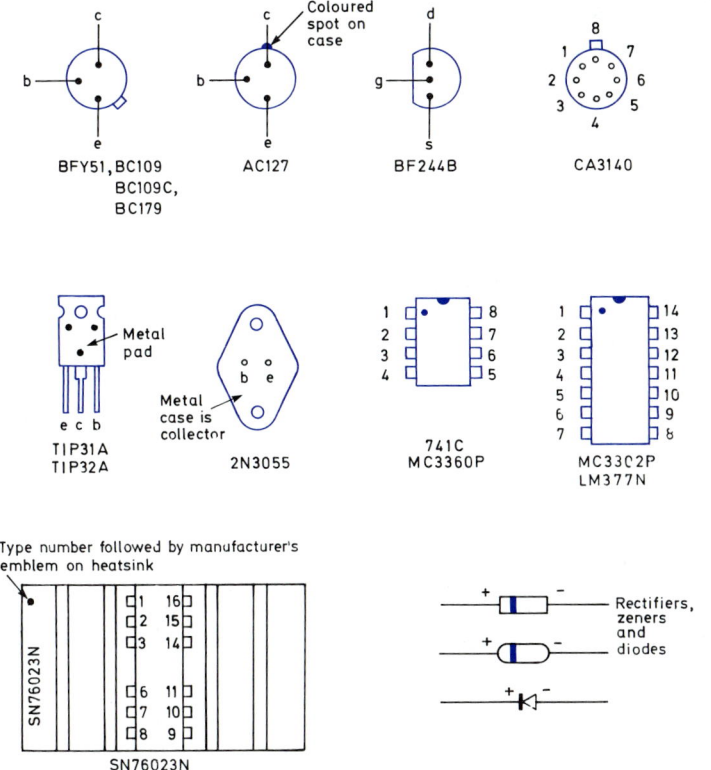

Figure A.1
Terminal identifications for semiconductors used in this book. Transistors show bottom views, i.c.s are top views

quite as easy. The same also applies to diodes and integrated circuits. These illustrations should clarify the conventions used for the terminals for all the semiconductors used in this book.

Transistors are bottom views, i.c.s are top views.

Resistor colour code

Most resistors are colour coded to indicate their resistance in ohms and the tolerance (the amount by which the actual value might differ from that marked).

There are four coloured bands around the body of the resistor and the first three give the resistance. The first coloured band (the one nearest the end of the body) corresponds to the first digit, the second colour gives the second digit and the third band tells you how many noughts follow these two digits.

The figures corresponding to the colours are:

black	0
brown	1
red	2
orange	3
yellow	4
green	5
blue	6
violet	7
grey	8
white	9

Yellow
Violet
Orange

4 7 3 (noughts)

= 47 000 ohms

The fourth band indicates the tolerance of the resistor. Only four colours are used here:

silver	10%
gold	5%
red	2%
brown	1%

Tolerance band

Absence of a fourth band indicates 20% tolerance.